EasyTerms™
Terminology Guidebook
for Zoology

Copyright 2009, Ed Creager

This edition of EasyTerms is one in a series of simple-to-use, college-level terminology guidebooks.

Although these guidebooks were originally intended for college students, many High School students will also find them helpful as they prepare for college.

Other topics covered in existing or forthcoming editions:

- Anatomy & Physiology (Human)
- Biochemistry
- Biology
- Botany
- Business Management
- Cell Biology
- Ecology
- Genetics
- Microbiology
- Nursing
- Psychology

EasyTerms can help support your educational advancement and can boost the vocabulary of almost anyone who reads it.

For more information on these and other publications, please visit the author's site:

AppleCreekBooks.weebly.com

and please note the author's "signature book" entitled,

"The Money-Saving Idea Book: Inside Tips for Starving Students, Frugal Seniors and Every Financial Survivor."

Foreword

This Zoology edition is a simple-to-use, college-level* terminology guidebook and is part of the EasyTerms reference series. In the book, terms are arranged alphabetically within appropriate topic areas. The complete index makes it easy to find any term and its definition.

* These books can also help High School students prepare so that, before they attend college, they'll already know a considerable amount of the terminology they'll need.

A substantial number of the terms defined here have additional definitions outside the scope of the subject being covered. More general definitions and additional meanings, if sought, are to be found in less specialized publications such as dictionaries and encyclopedias.

Please check the website of the author...

ApplecreekBooks.weebly.com

for more information on other available books.

To buy the Biology edition of EasyTerms and receive a Preferred Customer discount of 25%, please go to www.tinyurl.com/bookonbiology, click on "Add to Cart," and enter the **discount code ""** during check-out.

To save 25% on "The Money-Saving Idea Book," go to www.tinyurl.com/tmsib, click on "Add to Cart," and enter the **discount code "D7LM5VHK"** during check-out.

Important Notice:

EasyTerms™
Terminology Guidebook

Table of Contents

The terms that follow are divided into the topics shown below. The page number on which the topic begins is given. Within each topic, the terms are arranged alphabetically.

Introduction

1. abdomen

Body region between the diaphragm and the pelvis.

2. abdominal

Of the abdomen.

3. aboral

Away from the mouth.

4. anatomy

The study of structure.

5. anterior

Ventral, toward the belly.

6. axial

Concerning the axis or central plane that divides an animal body in halves.

7. bilateral

Relating to both sides of the body; having left and right sides.

8. biology

The study of life.

9. biosphere

An interconnected system over the earth's surface in which organisms exist.

10. community

A set of interacting organisms living in a given location.

11. **complementarity**

A relationship in which structures and their functions are mutually reinforcing.

12. **continuity of life**

The transfer of life from parent to offspring.

13. **corona**

Crown-like part.

14. **cranial**

Pertaining to the cranium.

15. **data**

Observations from an experiment.

16. **deductive reasoning**

Reasoning from a general statement to a specific case.

17. **development**

Process of increasing in complexity.

18. **distal**

Most distant from the point of origin of a structure.

19. **diversity**

The quality of variation among a class, such as living organisms.

20. **dorsal**

Toward the back; posterior in humans.

21. **ecology**

The study of how living organisms relate to each other and to their environment.

22. ecosystem

All the organisms in a natural setting and their physical environment.

23. environment

Changeable surroundings around a living organism.

24. excitability

Responsiveness to a stimulus.

25. external

On the outside.

26. feedback

The use of the output of a process to influence the process.

27. growth

Ability to increase in size.

28. homeostasis

The maintenance of a narrow, tolerable range of internal conditions.

29. homeostatic system

A control system that helps to maintain homeostasis.

30. hypothesis

A possible answer to a question; a possible explanation for observations that can be used to predict future outcomes.

31. inductive reasoning

Development of a general statement from a collection of observations.

32. internal

Inside or within.

33.   internal environment

The environment around cells, but within the body.

34.   interstitial

Concerning spaces between cells.

35.   lateral

On or toward the side.

36.   liter

The basic metric unit of fluid volume; 1.06 quarts.

37.   macromolecule

A very large molecule such as a protein or nucleic acid.

38.   mass

The amount of matter in an object.

39.   median

Toward or in the middle.

40.   meter

The basic metric unit of length; 39.37 inches.

41.   microscope

An instrument for observing structures too small to see with the naked eye.

42.   natural selection

A process by which well adapted organisms survive and reproduce in larger numbers than less well adapted ones.

43.   nucleus (of an atom)

Central part of an atom or a cell.

44. organ

A structure made up of several tissues that carries out particular functions; component of a system.

45. organ system

Group of organs that function together to carry out particular functions.

46. organism

A living thing.

47. organizational complexity

A concept that concerns the structural levels of an organism.

48. pectoral

Concerning the chest or breast.

49. pelvic

Concerning the hip or pelvis.

50. peripheral

Outer or away from the center.

51. physiology

The study of life functions.

52. polarity

Having positive or negative charge; having head and tail ends.

53. posterior

Toward the rear, dorsal in humans.

54. proximal

Nearest the point of origin of a structure.

55. reproduction

Process by which offspring arise.

56. responsiveness

Ability to react to a stimulus.

57. sagittal

A plane dividing right and left sides.

58. scientific method

A method of collecting and testing data pertaining to a scientific question.

59. systemic

Concerning the whole body.

60. theory

An explanation that accounts for many observations.

61. thorax

Chest; body region above diaphragm.

62. tissue

A group of similar cells, including intercellular substances, that carry out a particular function.

63. transverse

Crosswise.

64. ventral

Of or toward the belly.

65. viscera

Internal organs.

66. weight

The amount of gravitational force between two objects.

67. zoology

The study of animals.

Basic Chemistry

68.   acid

An ionizing substance that donates hydrogen ions.

69.   adenine

Purine found in nucleotides that pairs with thymine in DNA.

70.   alkaline

Basic, able to accept hydrogen ions.

71.   amino acid

A molecule having both acid and amino functional groups.

72.   anion

A negatively charged ion.

73.   atom

Smallest particle that retains properties of an element.

74.   atomic number

The number of protons in the nucleus of an atom.

75.   atomic orbital

Electron distribution around the nucleus of an atom.

76.   atomic weight

The total number of protons and neutrons in an atom; the average number if there are isotopes of the element.

77.   base

An ionizing substance that accepts hydrogen ions or reacts with an acid to form a salt.

78. buffer

A substance that resists pH change by holding or releasing hydrogen ions in a solution.

79. carbohydrate

An organic compound having several alcohol groups and an aldehyde or ketone group.

80. catalyst

A substance that increases a chemical reaction rate.

81. cation

A positively charged ion.

82. colloid

Glue-like; a particle in a colloidal dispersion.

83. colloidal dispersion

A state of matter with small particles suspended in a medium.

84. compound

A substance with two or more elements combined in definite proportion.

85. covalent bond

A chemical bond formed by shared electrons between two atoms.

86. dehydration

Removal of water.

87. denaturation

An alteration in the shape and properties of a protein molecule.

88. deoxyribonuclease

An enzyme that digests DNA.

89. deoxyribonucleic acid (DNA)

A nucleic acid in chromosomes that directs protein synthesis and transmits genetic information to a new generation.

90. deoxyribose

Five-carbon sugar found in DNA.

91. disaccharide

A molecule having two sugar (saccharide) units held together by a glycosidic bond.

92. electron

A negatively charged particle that continually moves around the nucleus of an atom.

93. element

A fundamental unit of matter.

94. endergonic

Requiring energy, as in a chemical reaction.

95. energy

The ability to do work.

96. entropy

Tendency toward chaos or disorder.

97. essential amino acid

Amino acid required in the diet because the body cannot make it.

98. essential fatty acid

Fatty acid required in the diet because the body cannot make it.

99. exergonic

Releasing energy, as in a chemical reaction.

100. fatty acid

A long hydrocarbon chain with a carboxyl group at one end.

101. functional group

A component of a molecule that participates in a chemical reaction.

102. glycine

An amino acid with the simplest chemical structure.

103. glycolipid

A molecule that contains both carbohydrate and lipid components.

104. glycoprotein

A molecule that contains both carbohydrate and protein components.

105. gram molecular weight

The quantity of a substance (in grams) equal to its molecular weight.

106. hydrogen bond

Weak covalent bond between hydrogen and another element, such as oxygen or nitrogen.

107. hydrolysis

The splitting of a molecule with the addition of water.

108. hydrophilic

Attacted to water.

109. hydrophobic

Tending to avoid water.

110. ion

A charged atom or group of atoms.

111. ionic bond

A chemical bond with atoms held together by the attraction of unlike charges.

112. isomer

A molecule having the same kinds and number of atoms as another molecule, but arranged differently.

113. isotope

An atom having a different number of neutrons than certain other atoms of the same element.

114. ketone body

Chemical substance that remains from incomplete digestion of fatty acids.

115. kinetic

Pertaining to the energy of motion.

116. lecithin

A phospholipid characteristic of animal tissues.

117. lipid

Fat or fatlike substance.

118. lipoprotein

A molecule made of lipid and protein.

119. mixture

Two or more substances combined in any proportions and retaining their individual properties.

120. mole

A gram molecular weight.

121. molecule

The smallest quantity of a substance that retains its chemical properties.

122. monomer

Simple molecule that can be linked to others to form a polymer.

123. monosaccharide

A simple sugar.

124. neutron

An uncharged particle in the nucleus of an atom.

125. nonpolar

Lacking charged regions.

126. nuclease

Enzyme that breaks down nucleic acids.

127. nucleic acid

A polymer of nucleotides; DNA or RNA.

128. nucleotide

A molecule having a nitrogenous base, a 5-carbon sugar, and one or more phosphates.

129. orbital

Electron distribution around the nucleus of an atom.

130. organic

Containing carbon.

131. oxidation

Addition of oxygen or loss of electrons in a chemical reaction.

132. peptide bond

A chemical bond between the amino group of one amino acid and the carboxyl group of another.

133. **pH**

The negative logarithm of the hydrogen ion concentration; a scale for expressing acidity or alkalinity.

134. **phospholipid**

A lipid made of glycerol, fatty acids, and phosphoric acid.

135. **polar compound**

A molecule having a charged area or polarity.

136. **polymer**

A molecule consisting of repeating units.

137. **polypeptide**

A chain of amino acids held together by peptide bonds.

138. **polysaccharide**

A molecule consisting of many saccharide units connected by glycosidic bonds.

139. **potential energy**

Energy due to position and capable of being released, as in a rock at the top of a hill.

140. **prosthetic group**

Cofactor bound to an enzyme.

141. **protease**

Enzyme that breaks down proteins.

142. **protein**

A polymer of amino acids.

143. **proteoglycan**

Molecule with a protein core and sugar molecules projecting from it.

144. proton

A positively charged particle in the nucleus of an atom.

145. radiation

Spreading from a center; giving off electromagnetic particles and waves.

146. reactant

A substance that enters into a chemical reaction.

147. reduction

Gain of an electron or loss of oxygen in a chemical reaction.

148. ribonuclease

Enzyme that digests RNA.

149. ribonucleic acid (RNA)

A nucleic acid made from information in DNA that is involved in protein synthesis.

150. salt

Chemical substance formed by replacement of hydrogen of an acid with another ion.

151. saturated fatty acid

A fatty acid lacking double bonds in the carbon chain and being saturated with hydrogen.

152. specific heat

The amount of heat needed to increase the temperature of a specific volume of substance one degree Celsius.

153. stereoisomer

Compound having the same kind and number of atoms as another compound, but in a different spatial arrangement.

154. steroid

A lipid with a complex four-ring structure.

155. **trace element**

A chemical element normally present in very small amounts in the body.

156. **triglyceride**

A triacylglycerol (glycerol and three fatty acids).

157. **unsaturated fatty acid**

Fatty acid with pairs of hydrogen atoms replaced by double bonds in the carbon chain.

158. **uridine triphosphate (UTP)**

A high energy molecule.

159. **valence**

An ion's charge.

Cells and Tissues

160. adipose

Pertaining to fat.

161. anaphase

A mitotic stage during which chromosomes move apart.

162. aster

Short microtubules at the ends of a spindle in a dividing animal cell.

163. basal body

Centriole at the base of a cilium or flagellum.

164. basal lamina

Thin layer of collagen fibers that support a tissue.

165. canaliculus

Tiny canal, as those between osteocytes in bone.

166. cell

A basic functional unit of a living organism.

167. cell cycle

A repetitive sequence of events involving DNA replication and cell division.

168. cell theory

A theory stating that living things are composed of cells.

169. centriole

One of a pair of intracellular bodies that participate in forming a mitotic spindle.

170. chondrocyte

Cell of the tissue cartilage.

171. chromatid

One of a pair of subunits in a duplicated chromosome.

172. chromatin

Nuclear material that condenses into distinct chromosomes during cell division.

173. chromosome

In a human cell, one of 46 nuclear structures made of DNA and protein.

174. cilium

A tiny hairlike projection found on some epithelial cells.

175. cisterna

Cavity in an organelle or organ.

176. collagen

A fibrous protein in connective tissue.

177. connective tissue

Tissue of fibrocytes and other cells imbedded in ground substance deposited in an organic fibrous matrix.

178. cytokinesis

Division of the cytoplasm that follows division of a nucleus.

179. cytoplasm

Cell substance, excluding the nucleus.

180. cytoplasmic streaming

Movement of chloroplasts and other structures within a cell's cytoplasm.

181. cytoskeleton

The organelles forming a cell's internal framework.

182. cytosol

The fluid part of cytoplasm that suspends organelles.

183. daughter chromosome

A chromosome produced in mitosis or second meiotic division as a result of replication and separation of strands.

184. diffusion

Movement of molecules from region of higher to lower concentration by kinetic energy.

185. duct

A tube that usually carries a secretion.

186. endoplasmic reticulum

A membranous vesicular network within a cell.

187. epidermis

Outer skin layer consisting of epithelium.

188. epithelium

A thin tissue that lines hollow organs or covers surfaces.

189. eukaryotic

Having a nucleus and membrane-bound organelles.

190. exocrine

Of a gland with ducts.

191. fibroblast

A connective tissue cell that makes fibers and ground substance.

192. flagellum

A movable hairlike process on a cell.

193. glycocalyx

Glycoprotein layer on the surface of a cell membrane.

194. Golgi apparatus

Membranous vesicles clustered in cells that complete synthesis of secretions.

195. haploid

Having one of a pair of chromosomes.

196. histology

The study of tissues.

197. intercalated

Inserted between other structures.

198. interphase

A cell cycle stage during which the cell is not dividing.

199. intracellular

Within a cell.

200. intrinsic

Entirely within.

201. karyokinesis

Division of a nucleus.

202. karyotype

Arrangement of chromosomes from a cell in pairs and in a fixed order.

203. kinetosome

A small body similar to a centriole at the base of cilia and flagella.

204. lacuna

Cavity in bone or cartilage usually containing one or more cells.

205. lysosome

Membrane-bound organelle that contains digestive enzymes.

206. malignancy

A tendency to become more virulent; a cancerous growth.

207. matrix

A fibrous framework in which ground substance of connective tissue is deposited.

208. metaphase

A stage in mitosis during which chromosomes align along the middle of a cell.

209. metaphasic chromosomes

Chromosomes which are in a condensed state during mitosis.

210. metastasis

The transfer of disease from one organ to another.

211. microfilament

A small, hollow protein fiber in cytoplasm that aids in movement or forms part of a cytoskeleton.

212. microtubule

A cylindrical organelle that forms part of a cell's mitotic spindle.

213. mitochondrion

An organelle that contains enzymes for oxidative and energy-capturing processes.

214. mitosis

Nuclear division that produces two identical nuclei.

215. mucus

Glycoprotein secretion from glands that lubricate surfaces.

216. muscle

A tissue capable of contracting.

217. nervous tissue

A tissue capable of conducting signals or impulses.

218. nucleolus

A body containing RNA within a nucleus.

219. nucleoplasm

The substance of a nucleus.

220. nucleus (of a cell)

Control center of a cell that contains genetic information.

221. organelle

A tiny function unit within a cell.

222. peroxisome

An organelle containing oxidative enzymes.

223. pinocytosis

Engulfing of small droplets.

224. prokaryotic

Lacking a nucleus and membrane-bound organelles.

225. prophase

The first mitotic stage during which the chromosomes become distinct.

226. protoplasm

Cell substance; literally, first formed.

227. remission

Abatement of disease symptoms or the period during which the abatement occurs.

228. ribosome

An organelle containing ribonucleic acid and protein where protein synthesis occurs.

229. saccule

Small hollow organ containing calcium carbonate stones that acts as receptor for static balance.

230. secretion

A cell product; the active transport of substances from the blood to the kidney filtrate.

231. sex chromosome

A chromosome associated with maleness or femaleness; X or Y chromosome in mammals.

232. sister chromatid

Half a doubled chromosome.

233. spindle fiber

Microtubules in eukaryotic cells involved in the movement of chromosomes during mitosis and meiosis.

234. stratum

A layer, usually of tissue.

235. telophase

The last mitotic stage during which nuclei reform.

236. teratogen

An agent that causes defective embryonic development.

237. tubule

Small tube.

238. tubulin

A protein that forms intracellular microtubules.

239. tumor necrosis factor

A substance that causes degeneration and death of tumor cells.

240. vacuole

Small space within a cell.

Movement Across Membranes

241. active transport

Transport of a substance against a gradient using a carrier molecule, enzyme, and cellular energy.

242. adsorptive endocytosis

Entry of a substance into a cell after attaching to the cell membrane.

243. cell membrane

Lipid and protein compounds that form the boundary of a cell.

244. endocytosis

Movement of fluid or particles across a membrane into a cell.

245. extracellular

Outside a cell.

246. facilitated diffusion

Diffusion down a gradient on a carrier molecule but not requiring cellular energy.

247. filtration

Passage of a fluid across a membrane by mechanical pressure.

248. fluid-mosaic model

A model of molecular arrangements in a cell membrane.

249. gel

A semi-solid state in a colloidal dispersion.

250. gradient

The rate of change in the magnitude of concentration, pressure, or other variable.

251. **hydrostatic pressure**

Force exerted by a fluid.

252. **hyperosmotic**

Having higher osmotic pressure than a reference solution.

253. **hypertonic**

Causing movement of water out of cells.

254. **hyposmotic**

Having lower osmotic pressure than a reference solution.

255. **hypotonic**

Causing movement of water into cells.

256. **isosmotic**

Having the same osmotic pressure as a reference solution.

257. **isotonic**

Causing no net water movement across a cell membrane.

258. **osmolarity**

A solution's osmotic concentration determined by the number of osmotically active particles it contains.

259. **osmosis**

Diffusion of water through a membrane from its own higher to a lower concentration.

260. **osmotic pressure**

Pressure created by osmosis.

261. **passive transport**

A process that moves substances without energy expenditure by the organism.

262. permeability

Membrane property that allows a substance to cross it.

263. plasma membrane

Membrane forming the boundary of a cell.

264. plasmolysis

Contraction of cytoplasm because of water loss.

265. selectively permeable

A membrane property that allows passage of some substances while preventing passage of others.

266. sodium-potassium pump

Mechanism that actively moves Na ions out of cells and K ions into them against gradients.

267. sol

A liquid state of a colloidal dispersion.

268. solute

A dissolved substance.

269. solution

A liquid containing dissolved substances.

270. solvent

A substance in which other substances can dissolve.

271. surface tension

Resistance to rupture by the surface film of a liquid.

272. surface-to-volume ratio

The surface area of a structure divided by its volume.

273. tonicity

The degree to which fluid can move into or out of cells.

Cellular Metabolism

274. activation energy

Energy needed to start a chemical reaction.

275. adenosine triphosphate

An important energy storage molecule.

276. aerobic

In the presence of oxygen.

277. alcoholism

Disease of repetitive, excessive alcohol intake, which leads to severe problems in daily living.

278. amylase

An enzyme that digests starch.

279. anabolic

Of anabolism.

280. anabolism

Synthetic, energy using process.

281. anaerobic

Lacking oxygen.

282. anorexia nervosa

A serious neurological disorder in which a person loses weight and becomes emaciated.

283. apoenzyme

Protein part of an enzyme that requires a coenzyme to become functional.

284. **basal metabolic rate (BMR)**

Amount of energy used to maintain life in an awake, resting individual.

285. **basal metabolism**

The process of using energy from nutrients to maintain life in an awake, resting state.

286. **binding site**

A site where a particular molecule binds to a membrane or other structure.

287. **bioassay**

Measuring concentration or effect of substance by observing its effect on a living organism.

288. **biological oxidation**

Transfer of electrons in a mitochondrion and concurrent capture of energy in ATP.

289. **bulimia**

Binge eating, usually followed by self-induced vomiting.

290. **calorie**

Quantity of heat needed to raise the temperature of one gram water one degree Celsius.

291. **catabolism**

Breakdown of molecules that makes energy available.

292. **centromere**

Site on a chromosome to which spindle fiber attaches.

293. **chemosynthesis**

Production of organic molecules from inorganic molecules and elements.

294. **coenzyme**

A substance that works with an enzyme in activating chemical reactions.

295. cofactor

Small molecule or ion that helps to activate an enzyme.

296. core body temperature

Temperature deep within the body.

297. cytochrome

Enzyme that participates in electron transport and energy capture in ATP.

298. deamination

Removal of an amino group.

299. decarboxylation

Removal of a carboxyl (-COOH) group form a molecule.

300. dehydrogenation

Oxidation by the removal of hydrogen atoms from a molecule.

301. detoxification

Enzyme controlled reaction that reduced toxicity of a substance.

302. diabetes mellitus

A disorder due to lack or inactivity of insulin that allows glucose to accumulate in the blood and urine.

303. electron transport system

Enzymes and coenzymes in cristae of mitochondria that move electrons from substrates to oxygen.

304. endopeptidase

Enzyme that breaks peptide bonds within a peptide chain.

305. endotherm

An animal whose body temperature is maintained in a narrow range by internal processes.

306. enzyme

A protein that increases the rate of a chemical reaction in a living organism.

307. exopeptidase

Enzyme that breaks peptide bonds at ends of peptide chains.

308. fermentation

An anaerobic metabolic process in which carbohydrate is broken down to alcohol and other simple molecules.

309. fever

An abnormally high body temperature.

310. first law of thermodynamics

Total energy in a system remains constant but can be converted from one form to another.

311. flavin adenine dinucleotide (FAD)

A coenzyme that carries hydrogen.

312. fluorescence

Emission of light by a substance that has absorbed light of a different wave length.

313. free energy

Energy available to do work.

314. gluconeogenesis

Metabolic pathway that makes glucose from noncarbohydrate substances.

315. glycogenesis

Metabolic pathway for glycogen synthesis.

316. glycogenolysis

Metabolic pathway for glycogen breakdown.

317. glycolysis

Metabolic pathway for breakdown of glucose to pyruvic acid.

318. heterotherm

Animal whose body temperature fluctuates with environmental temperature.

319. high density lipoprotein (HDL)

A blood particle containing protein and lipid that tends not to deposit cholesterol in blood vessels.

320. homeotherm

Animal that maintains a constant body temperature.

321. hyperthermia

An abnormally high body temperature; fever.

322. hypothermia

An abnormally low body temperature.

323. kilocalorie

Heat required to raise the temperature of one kilogram of water one degree Celsius.

324. Krebs cycle

Metabolic pathway that oxidizes acetyl-CoA; citric acid cycle; tricarboxylic acid cycle.

325. ligand

That which binds to a receptor.

326. low density lipoprotein (LDL)

A blood particle containing protein and lipic implicated in the deposition of cholesterol in blood vessel walls.

327. lysis

Disintegration, usually by enzyme action.

328. **metabolic rate**

Rate at which nutrients are oxidized.

329. **metabolism**

All chemical reactions in a living organism.

330. **net protein utilization**

Proportion of protein eaten that is actually used by cells.

331. **nicotinamide adenine dinucleotide (NAD)**

A coenzyme that transports hydrogen atoms or electrons in oxidation-reduction reactions.

332. **obesity**

An excessive amount of fat.

333. **oxidative phosphorylation**

Capture of energy in ATP during oxidative metabolism.

334. **phosphorescence**

Light emission without heat.

335. **phosphorylation**

Binding of a phosphate group to a molecule.

336. **phototrophic**

Producing organic molecules by using light energy.

337. **poikilothermic**

Having a body temperature that fluctuates with that of the environment.

338. **putrefaction**

Enzymatic anaerobic degradation of proteins.

339. pyrogen

A substance that causes fever.

340. reabsorption

Return to the body of a substance that had been secreted.

341. receptor

A specific site with which a specific substance can bind; cell that responds to signals sensed from the environment.

342. saturation

Condition of having all chemical affinities satisfied.

343. second law of thermodynamics

Natural processes increase in randomness and dissipate energy.

344. specificity

The attribute of being specific.

345. substrate (of an enzyme)

Substance on which an enzyme acts.

346. thermogenesis

Heat generation.

347. total parental nutrition (TPN)

Process of giving all required nutrients by a route other than the digestive tract.

348. transamination

Transfer of an amino group from one molecule to another.

349. turnover

Reuse of a substance made available by a catabolic reaction.

350.   very low density lipoprotein (VLDL)

A particle in the blood containing much lipid and little protein.

Body Form and Support

351. appendage

A movable extension of the body, such as a leg or arm.

352. auditory vesicle

A small bone of the middle ear that transmits vibrations.

353. bilateral symmetry

Symmetry in which a plane can divide equal left and right halves.

354. biradial symmetry

Having two planes of symmetry at right angles to each other.

355. brachial

Of the arm.

356. bursa

A sac containing synovial fluid located at pressure points or near joints.

357. calcification

Depositing of calcium salts in an organic matrix.

358. callus

A thickened area.

359. carpal

A wrist bone.

360. cartilage

A firm, resilient, flexible connective tissue.

361. clavicle

Bone that extends from the sternum to the scapula.

362. concha

One of several shell-shaped bones in the nasal cavity.

363. coxal bone

A bone that forms half of the pelvic girdle.

364. cranium

Skull bones that surround the brain.

365. cutaneous

Pertaining to skin.

366. cuticle

The outer layer of skin, especially around nails.

367. dermal bone

Bone formed directly in connective tissue membrane.

368. dermis

A thick skin layer underlying the epidermis.

369. desmosome

Discontinuous sites of attachment between adjacent cells.

370. diaphragm

A thin partition, as is formed by a muscle between the thoracic and abdominal cavities.

371. eleidin

A keratin precursor found in the stratum lucidum.

372. endoskeleton

An inner skeleton.

373. exoskeleton

An external skeleton.

374. fibula

Lateral leg bone found between the knee and ankle.

375. fontanel

A membranous non-bony region between cranial bones in an infant.

376. foramen

Opening in a body structure, such as a bone.

377. fracture

A break, such as in a bone.

378. frontal

Of the forehead.

379. fulcrum

Fixed point about which a lever produces movement.

380. haversian canal

Passageway in a bone through which blood vessels and nerves pass.

381. humerus

The bone of the upper arm.

382. hydroskeleton

Turgidity of fluid in a body space that supports the organism.

383. **ilium**

The posteriolateral bone of the pelvis.

384. **incus**

Anvil; the ear bone that receives vibrations from the malleus.

385. **integument**

Outer covering; skin.

386. **invagination**

Infolding of a part.

387. **ischium**

A posterior bone of the pelvic girdle.

388. **keratin**

Water-insoluble protein in epidermis of vertebrates.

389. **lamella**

Layer or plate.

390. **lorica**

Loose case made of secreted matter and detritus.

391. **malleus**

Hammer; the outermost bone of the middle ear.

392. **marrow**

Fatty substance found in a marrow cavity.

393. **maxilla**

The facial bone that contains sockets for upper teeth.

394. mesenchyme

Embryonic mesoderm.

395. metacarpal

One of five bones in the palm of the hand.

396. metatarsal

One of five long bones in the foot.

397. occipital

Of or near the back of the head.

398. ossification

Mineral deposition in the process of bone formation.

399. osteon

The cells, matrix, and passages that make up a unit of compact bone.

400. ostium

Opening to or from a tube or cavity.

401. patella

A small bone that forms the kneecap.

402. pedicel

Stalk.

403. phalanges

Small, slightly elongated bones of the fingers and toes.

404. plantar

Of the sole of the foot.

405. plasticity

Able to be molded or formed; changeability.

406. pubis

An anterior bone of the pelvic girdle.

407. radial symmetry

Symmetry in which all wedge-shaped sections around a vertical line are similar.

408. radius

The smaller of the forearm bones.

409. raphe

A seamlike ridge, usually where two structures have fused.

410. rickets

A failure of bones to harden in childhood because of a calcium deficiency.

411. sacrum

A bone formed from the fusion of five vertebrae that articulates posteriorly with the pelvic girdle.

412. scapula

A large bone lateral to the vertebrae in the shoulder area; shoulder blade in humans.

413. sclerotome

An embryonic tissue from which vertebrae and ribs develop.

414. sebum

A substance containing oils and epithelial cell debris from sebaceous glands.

415. sedentary

Remaining in one place.

416. septum

A wall or partition.

417. sinus

A cavity or recess.

418. smooth muscle

A type of muscle located in the walls of hollow organs and blood vessels.

419. spicule

A needle-shaped structure.

420. stapes

The middle ear bone that transmits vibrations to the oval window; stirrup.

421. sulcus

A furrow or groove.

422. suture

A fibrous joint at which no movement occurs.

423. symmetry

Similarity of form or arrangement around a point, line, or plane.

424. symphysis

A cartilaginous, slightly moveable joint.

425. syncytium

A group of cells that lack membranes to separate them.

426. synovial joint

A freely movable joint.

427. **tarsal**

Of the foot bones.

428. **temporal**

Time-related; a brain lobe where auditory and olfactory areas are located.

429. **tentacle**

Long, nonsegmented protrusion from body wall.

430. **tibia**

Larger, more medial weight-bearing bone of the leg.

431. **trochlea**

Pulley.

432. **ulna**

Larger of the forearm bones.

433. **vertebra**

Back bone.

434. **vomer**

A shovel-shaped bone that forms the nasal septum.

435. **zygomatic**

Of the cheek bone.

Movement

436. A band

Band equivalent to the length of a myosin molecule.

437. abduction

Motion of a body part away from the midline.

438. Achilles tendon

A tendon in the heel.

439. actin

A contractile protein.

440. action

A movement produced by one or more muscles.

441. adductor

Muscle that draws a structure toward the midline.

442. agonist

The prime mover among a group of muscles.

443. ameboid motion

Cellular movement by protoplasmic streaming.

444. amoebocyte

Somatic cell that can move by amoeboid movement.

445. antagonist

A muscle that opposes an angonist.

446. apodeme

Tendon-like extension of cuticle to which muscles attach in arthropod joints.

447. articulation

A joint.

448. atrophy

A decrease in size, usually accompanied by reduced function.

449. autotomy

Self-amputation, as of an appendage that can be regenerated.

450. ball-and-socket joint

A joint with a ball-shaped articular surface of one bone fitting into a socket-shaped articular surface of another.

451. bipedal locomotion

Moving on two legs.

452. brachiation

Arboreal locomotion by swinging by the arms.

453. chemotropism

Growth in response to a chemical stimulus.

454. contractile protein

A protein that acts in shortening a muscle or causing it to develop tension.

455. contractility

The ability to develop tension or shorten.

456. contraction cycle

Repetitive sliding actions of actin and myosin in a muscle filament as it develops tension.

457. creatine phosphate

A molecule that accounts for limited energy storage in muscle.

458. cross-bridge

The end of a myosin filament bound to actin during muscle contraction.

459. effector

Structure by which an organism acts; muscle, gland, cilium.

460. excitation-contraction coupling

The means by which neural signals excite muscle cells and cause contraction.

461. extensibility

Ability to be extended or stretched.

462. extensor

Muscle that extends or straightens a limb.

463. fascia

Fibrous connective tissue sheath around muscles and beneath skin.

464. fatigue

Loss of power for a short time.

465. flexion

A movement that decreases the angle between two bones.

466. hydraulic movement

Movement achieved by fluid manipulation in a hydrostatic skeleton.

467. I band

Isotropic band formed mainly by actin molecules in a striated muscle.

468. insertion

The most moveable attachment of a muscle to a bone.

469. isometric

Having the same length.

470. joint

A connection between two or more bones.

471. kinesis

Activity of an organism in response to a stimulus.

472. latent period

In muscle physiology, a time between the application of a stimulus and the beginning of muscle contraction.

473. ligament

Cord of fibrous connective tissue that attaches bones to each other.

474. metachronal rhythm

Wavelike, coordinated beating of cilia.

475. mobility

Ability to move from place to place.

476. motor end plate

A portion of sarcolemma lying beneath nerve endings.

477. motor unit

A motor neuron and the muscle fibers it innervates.

478. myofibril

A contractile fiber in a muscle cell.

479. myofilament

A component of a myofibril consisting of one or more protein molecules.

480. myogenic

Contraction generated by specialized muscle cells.

481. myoglobin

A pigmented protein that binds oxygen in muscle tissue.

482. myomere

Segment of an animal that gives rise to muscles.

483. myoneural junction

The structure at which nerve and muscle tissues meet and impulses are relayed.

484. myosin

A protein that comprises thick filaments of a myofibril.

485. myotome

A block of mesoderm from which muscle arises.

486. neurogenic

Contraction generated by nerve impulses.

487. neuromuscular

Concerning the association between the nervous and muscular systems.

488. origin

The least movable attachment of a muscle to a bone.

489. oxygen debt

The quantity of oxygen required to oxidize metabolites produced anaerobically during strenuous activity.

490. paramyosin

Contractile protein found in certain muscles of mollusks and a few other invertebrates.

491. prime mover

A muscle that is the most direct cause of a particular movement.

492. pseudopodium

Footlike cytoplasmic extension usually used in locomotion.

493. rotation

Motion of a part about its own axis.

494. sarcolemma

Cell membrane of a muscle cell.

495. sarcomere

The contractile unit of skeletal muscle.

496. sarcoplasm

The protoplasmic, nonfibrillar substance of a muscle cell.

497. sarcoplasmic reticulum

A vesicular network associated with myofibrils of a striated muscle cell.

498. seta

A bristle that serves in locomotion of segmented worms.

499. sliding filament theory

An explanation of how myofilaments move with respect to each other during muscle contraction.

500. striation

Stripe.

501. summation

Addition, as in effects of multiple stimuli to a muscle.

502. synergist

A muscle that works with a prime mover.

503. tendon

A fibrous connective tissue cord that holds a muscle to a bone.

504. tension

A pulling force.

505. tetanus

A sustained contraction maintained by repeated muscle stimulation.

506. thigmotropism

Change in orientation in response to touch.

507. tonus

A slight continuous muscle contraction.

508. transverse (T) tubule

Crosswise tubule in skeletal muscle myofibrils that carries signals from the sarcolemma to the myofibrils.

509. tropomyosin

A muscle protein that alters the actin configuration so that contraction can occur.

510. troponin

A muscle protein that binds to tropomyosin causing it to alter the configuration of actin.

511. twitch

A muscle response to a single stimulus.

Digestive Structures

512. adventitia

Outermost connective tissue layer on an organ or blood vessel.

513. anus

An opening through which wastes exit the digestive tract.

514. argentaffin cell

Cell in the stomach lining that secretes histamine and serotonin.

515. bicuspid

Having two points or cusps.

516. bolus

A mass.

517. cecum

Blind pouch.

518. cementum

Material surrounding dentin in the root of a tooth.

519. chylomicron

A particle consisting of lipids and proteins made in the intestinal mucosa and released into lacteals.

520. colon

Large intestine from the cecum to the rectum.

521. cuspid

A point or tapering projection.

522. deciduous tooth

Tooth that is not permanent; 'baby tooth'.

523. dentin

Bonelike substance of a tooth located beneath the surface.

524. duodenum

A short part of the small intestine adjacent to stomach that receives secretions from the liver and pancreas.

525. enamel

Hard covering of a tooth seen above the gumline.

526. epiglottis

Elastic cartilage that closes the glottis.

527. esophagus

Muscular tube between the pharynx and stomach.

528. fundus

Part of an organ farthest from its outlet.

529. gallbladder

A sac on the underside of the liver where bile is stored.

530. gastric

Of the stomach.

531. gastrodermis

A layer of cells that lines the body cavity in cnidarians.

532. gastrovascular cavity

A central digestive cavity with a single opening found in lower animals.

533. gingiva

Gum.

534. ileum

Lower part of the small intestine.

535. incisor

A cutting tooth.

536. incomplete gut

Digestive tract without an anus.

537. jejunum

Middle part of the small intestine.

538. Kupffer's cell

A phagocytic cell in the walls of liver sinusoids.

539. lacteal

Lymph vessel in a villus of the small intestine.

540. mesentery

A two-layered membrane continuous with the peritoneum that suspends an abdominal organ.

541. micelle

A small fat droplet in chyme.

542. microvillus

A cytoplasmic projection of surface membrane of intestinal epithelial cells.

543. molar

A large grinding tooth; pertaining to the concentration of a solution.

544. mucosa

Mucous membrane lining cavities and passageways.

545. pancreas

A digestive gland that secretes enzymes and hormones.

546. peritoneum

A membrane that covers abdominal organs and lines the abdominal cavity.

547. Peyer's patch

An elevated lymphoid tissue mass in the mucosa of the small intestine.

548. pulp cavity

A chamber within a tooth that contains blood vessels and nerves.

549. pylorus

Stomach region attached to the small intestine.

550. rectum

Terminal portion of the digestive tract between the colon and the anal canal.

551. salivary gland

Gland that secretes saliva.

552. sphincter

A ringlike muscle by which a natural orifice opens and closes.

553. stomodeum

Ectodermal evagination from which the mouth and adjacent pharynx form.

554. submucosa

A layer beneath the intestinal mucosa.

555. villus

Vascular tuft.

Nutrition

556. absorption

Movement of substances across a membrane.

557. absorptive

Concerning absorption.

558. appendicitis

Inflammation of the appendix.

559. ascorbic acid

Vitamin C.

560. assimilation

Absorption and entry into cellular metabolism of products of digestion.

561. autotroph

An organism that makes organic compounds from inorganic substances in the environment.

562. bile

Liver secretion that aids in digestion by emulsifying fats.

563. carnivore

An animal whose diet consists mainly of other animals.

564. carotene

Yellow substance that usually has vitamin A activity.

565. CCK (cholecystokinin-pancreozymin)

An enteric hormone that stimulates release of bile from the gallbladder and enzymes from the pancreas.

566. cellulose

A polysaccharide found in the structure of many plants.

567. chyme

Semiliquid, partially digested food leaving the stomach.

568. chymotrypsin

A proteolytic enzyme from the pancreas.

569. cobalamin

Substance needed to make red blood cells; vitamin B-12.

570. coprophagy

Eating of feces.

571. defecation

Elimination of undigested refuse.

572. deglutition

Swallowing.

573. diarrhea

Excessively frequent, fluid bowel movements.

574. digestion

Breakdown of large molecules into smaller ones.

575. emulsification

Process by which bile salts cause fat droplets from foods to break into smaller particles.

576. enterocoel

Outpouching of primitive gut that forms a body cavity.

577. evaporation

Changing of a substance from liquid to gaseous form.

578. feces

Digestive waste expelled from the rectum through the anus.

579. folacin

Vitamin needed to help transfer single carbon groups.

580. gluten

A protein containing gliadin found in wheat and some other grains.

581. hepatic

Of the liver.

582. herbivore

An animal whose diet consists mainly of plants.

583. heterotroph

An organism that metabolizes ready made organic matter.

584. ingestion

Intake of food or fluid.

585. insectivorous

Pertaining to a plant that increases its nitrogen intake by capturing and ingesting insects.

586. intrinsic factor

Substance secreted by the gastric mucosa required for the transport and absorption of vitamin B12.

587. lactase

An enzyme that digests lactose.

588. lipase

An enzyme that breaks down lipids.

589. lumen

Cavity in a tubular organ.

590. macronutrient

Nutrient needed in relatively large amounts.

591. malnutrition

Ill health caused by an inadequate diet.

592. maltase

Enzyme that digests maltose, a disaccharide derived from starch.

593. micronutrient

A nutrient needed in relatively small quantities.

594. mineral

Inorganic substance.

595. niacin

B vitamin used to synthesize the coenzyme NAD.

596. nutrient

Any substance needed or usable in an organism's metabolic processes.

597. nutrition

The act of providing substances needed for good health through food ingestion.

598. omnivore

An animal that eats both plant and animal tissue.

599. pantothenic acid

B vitamin used to synthesize coenzyme A.

600. pepsin

An enzyme that starts breakdown of protein in the stomach.

601. peristalsis

Wavelike, propelling contractions along tubular passageways.

602. rennin

Enzyme from gastric mucosa of immature mammals that digests milk protein.

603. riboflavin

Heat-labile B vitamin used to synthesize the coenzyme FAD.

604. ruminant

A mammal that chews cud.

605. saprophyte

Organism that obtains nutrients from nonliving organic matter.

606. saprozoic

Concerning organisms that absorb nutrients from environment.

607. scurvy

A disease due to a vitamin C deficiency.

608. secretin

A hormone from the intestinal mucosa that stimulates secretion of bile and pancreatic fluid.

609. sucrase

An enzyme that digests sucrose.

610. thiamine

A water-soluble B vitamin used to synthesize cocarboxylase.

611. tocopherol

A substance with vitamin E activity.

612. trypsin

A proteolytic enzyme released from the pancreas.

613. typhlosole

Infolding that increases intestinal absorptive area.

614. vitamin A

Vitamin needed to synthesize visual pigments and maintain epithelial cells.

615. vitamin D

A vitamin that facilitates calcium absorption.

616. vitamin E

Vitamin that acts as an antioxidant.

617. vitamin K

Vitamin needed for synthesis of some blood clotting factors.

618. zymogen

Inactive enzyme.

Gas Exchange

619.   **acinus**

A small cluster.

620.   **Adam's apple**

Thyroid cartilage of larynx, which is prominent in males.

621.   **alveolus**

One of many small air sacs in the lungs and in secretory parts of some glands; a tooth socket.

622.   **apical**

Located at the apex or tip of a structure.

623.   **asthma**

A disorder in which constriction of bronchioles causes difficulty in breathing.

624.   **book gill**

A respiratory organ found in certain crabs.

625.   **book lung**

A respiratory organ found in spiders.

626.   **Boyle's law**

Pressure exerted by a gas is inversely proportional to its volume.

627.   **branchial**

Pertaining to gills.

628.   **bronchiole**

One of many small tubes in the lungs.

629. bronchitis

Inflammation of bronchi.

630. bronchus

One of many tubes leading from the trachea to the lung.

631. cardiopulmonary resuscitation (CPR)

A method for maintaining blood flow and gas exchange in a person with no heart beat and no breathing.

632. Dalton's law

Each gas in a mixture exerts a partial pressure that is independent of other gases.

633. emphysema

A disorder characterized by destruction or dilation of walls of alveoli.

634. Eustachian tube

A passage that connects the middle ear and the pharynx.

635. expiration

Exhaling; breathing out.

636. gas exchange

The diffusion of gases across membranes as when oxygen enters and carbon dioxide leaves blood.

637. gill

An aquatic animal's respiratory organ.

638. gill slit

Opening from pharynx to environment through which water leaves gill in fish.

639. glottis

A slitlike opening from the pharynx to the larynx.

640. Henry's law

A gas dissolves in a liquid in proportion to its solubility and its partial pressure.

641. inspiration

Breathing in.

642. larynx

Voice box.

643. lung

Internal sac having moist surfaces for gas exchange.

644. mantle cavity

A space between the body and the mantle of a mollusk; contains the gills.

645. nares

Openings to nasal cavities.

646. operculum

A protective covering over the gill chamber of a fish.

647. palate

Flat plate-like roof of the mouth.

648. partial pressure

Pressure exerted by one gas in a mixture of gases.

649. pharynx

Throat.

650. pleura

Membrane surrounding a lung and lining its cavity.

651. **pressure**

Force due to compression.

652. **pulmonary**

Of the lungs or the blood vessels carrying blood to and from gas exchange membranes.

653. **respiration**

The processes of ventilation (breathing) and gas exchange.

654. **respiratory center**

A neural center in the brain stem that regulates respiration.

655. **respiratory pigment**

A cytochrome or other pigment that increases the oxygen carrying capacity of blood or another tissue.

656. **spiracle**

A respiratory passageway in an arthropod.

657. **spirometry**

Measurement of gas volumes entering or leaving the lungs.

658. **surfactant**

A phospholipid that reduces surface tension.

659. **swim bladder**

A dorsal sac in the body cavity of a fish that allows it to regulate its bouyancy.

660. **tetrapod**

Having four appendages.

661. **tonsil**

An aggregate of pharyngeal lymphatic tissue.

662. **torr**

A unit of pressure equal to that required to support a column of mercury 1 mm tall.

663. **trachea**

Passage from the larynx to the bronchi.

664. **ventilation**

Movement of gases between the lungs and the environment.

665. **vital capacity**

The largest gas volume that can be expired after a maximal inspiration.

666. **vocal**

Of the voice.

Transport

667. **adhesion**

Tendency of unlike molecules to stick together.

668. **aneurysm**

A saclike dilation in an arterial wall.

669. **aorta**

A large artery that carries blood from the left ventricle to other arteries of the systemic circulation.

670. **aplastic**

Having no tendency to undergo cell division.

671. **arteriole**

A blood vessel between an artery and capillaries.

672. **artery**

A large blood vessel carrying blood away from the heart.

673. **artificial pacemaker**

A device that automatically stimulates the heart and maintains a regular heart rate.

674. **atrium**

A chamber or entrance.

675. **auricle**

An earlike appendage.

676. **Bohr effect**

Tendency of high oxygen concentration in the lungs to cause hemoglobin to release carbon dioxide.

677. capillary

A small blood vessel interposed between an arteriole and a venule.

678. carbonic anhydrase

Enzyme that catalyzes reversible reaction between carbonic acid and carbon dioxide and water.

679. cardiac

Of the heart.

680. cardiovascular

Of the heart and blood vessels.

681. celiac

Related to the abdomen.

682. circulation

The continuous passage of blood through blood vessels from one region of the body to another.

683. closed circulatory system

A continuous system of blood vessels that contain an animal's blood.

684. cohesion

The attraction of water molecules for each other.

685. conduction system

Fibers in heart muscle in which signals coordinate atrial and ventricular contractions.

686. conus arteriosus

A large artery in frogs through which blood leaves the heart.

687. diastole

Relaxation, as of heart muscle.

688.   dorsal root ganglion

A ganglion containing cell bodies of neurons that carry sensory signals.

689.   ductus arteriosus

Vessel derived from sixth aortic arch that connects aorta and pulmonary artery in embryos.

690.   edema

Excess fluid accumulation in the tissues.

691.   electrocardiogram (ECG)

A record of electrical changes detected on the body surface associated with heart contractions.

692.   electroencephalogram (EEG)

A record of electrical changes detected on the scalp and associated with brain activity.

693.   embolus

Particle carried by bloodstream that blocks a blood vessel.

694.   endocardium

The heart's epithelial lining.

695.   heart failure

Loss of ability to pump sufficient blood to supply body tissues with nutrients and remove wastes.

696.   heart murmur

An abnormal sound caused by turbulence around a defective valve.

697.   hypertension

Excessively high blood pressure.

698.   hypotension

Excessively low blood pressure.

699. ischemia

Reduction in blood flow to an area.

700. isovolumetric

Having the same volume.

701. mantle

A membranous organ that secretes the shell of a mollusk.

702. myocardium

The thick muscular layer of the heart.

703. oncotic pressure

Osmotic pressure created by the proteins and other molecules in a fluid.

704. open circulatory system

A system of blood vessels that carries blood between the heart and body cavities.

705. pacemaker

An aggregation of cells that spontaneously excite other cells, as in the sinoatrial node.

706. pericardium

A sac around the heart.

707. peripheral resistance

Degree of constriction of blood vessels that contributes to blood pressure by resisting flow.

708. plaque

A sheetlike deposit.

709. portal system

Set of veins that carry blood from one capillary bed to another.

710. pulse

Rhythmic expansion and contraction of an artery caused by the pumping action of the heart.

711. Purkinje fiber

The terminal ends of fibers in the heart's conduction system.

712. resistance

Opposition to flow, as in blood vessels.

713. SA node

Part of the heart's conduction system that normally initiates contractions.

714. sinoatrial node

Mass of tissue at union of vena cava and right atrium that regulates heart rate.

715. sinus venosus

First heart chamber in lower vertebrates.

716. sphygmomanometer

A device used to measure blood pressure.

717. stenosis

Narrowing or constriction.

718. stethoscope

An device used to amplify sounds from inside the body.

719. stroke volume

Volume of blood ejected by one ventricle during a single contraction.

720. systole

Contraction.

721. T wave

A part of an electrocardiogram that occurs as ventricles repolarize.

722. tricuspid

Having three points or cusps.

723. tunica

A layer or coat.

724. valve

A structure in a passageway that prevents reflux.

725. vascular

Pertaining to or full of vessels.

726. vascular system

A transport system.

727. vein

A blood vessel that carries blood to the heart.

728. vena cava

A large vein that empties blood into the heart.

729. venous

Of a vein.

730. ventricle

A small cavity.

731. venule

A small vessel between capillaries and a vein.

732. viscosity

A fluid's tendency to resist flow.

733. water vascular system

Hydraulic system in echinoderms that accomplishes gas exchange, food distribution, and locomotion.

Blood / Circulating Fluids

734.   agranular leukocyte

A white blood cell lacking cytoplasmic granules.

735.   albumin

A small protein made in the liver and released into blood.

736.   anemia

A hemoglobin deficiency associated with too few erythrocytes or poorly functioning ones.

737.   anticoagulant

A substances the prevents blood clotting.

738.   basophil

A leukocyte having granular cytoplasm and able to be stained with a basic dye.

739.   blood

Fluid pumped by the heart through a closed system of vessels.

740.   carbaminohemoglobin

Hemoglobin to which carbon dioxide is bound.

741.   cross-matching

Comparison of donor and prospective recipient bloods to detect possibilities of agglutination.

742.   diapedesis

Squeezing of leukocytes between the cells of capillary walls.

743.   eosinophil

A granular leukocyte capable of being stained with the dye eosin.

744. erythrocyte

Red blood cell.

745. erythropoiesis

Process of red blood cell formation.

746. erythropoietin

A substance from the kidney that stimulates erythropoiesis.

747. ferritin

A molecule made up of the protein apoferritin and iron.

748. fibrin

A fibrous protein that forms a network in a blood clot.

749. fibrinogen

Inactive fibrin.

750. globin

A globular protein found in hemoglobin and certain other biological molecules.

751. globulin

A globular shaped protein, many of which are found in plasma.

752. granular leukocyte

A white blood cell with granular cytoplasm.

753. hematocrit

The proportion of erythrocytes in a volume of blood.

754. hematopoiesis

The formation of blood cells.

755. heme

An iron-containing pigment in hemoglobin that binds oxygen.

756. hemodialysis

The removal of substances from the blood by dialysis.

757. hemoglobin

The oxygen-carrying protein in erythrocytes.

758. hemolysis

Breakdown of erythrocytes with hemoglobin release.

759. hemophilia

An inherited inability to produce a blood clotting factor.

760. hemopoiesis

The formation of blood cells.

761. hemorrhage

Loss of a significant volume of blood.

762. hemostasis

The arrest of bleeding.

763. hirudin

An anticoagulant secreted by leeches.

764. jaundice

Yellowish tone to skin and membranes caused by excess bile pigments in the blood.

765. leukocyte

A white blood cell.

766. lymph

Interstitial fluid in a lymphatic vessel.

767. lymph node

A lymphatic tissue aggregation interposed at intervals along a lymphatic vessel.

768. lymphatic

Concerning lymph or a vessel that carries it.

769. lymphocyte

A leukocyte that participates in an immune response.

770. lymphoid

Resembling lymph or lymphatic tissue.

771. monocyte

A large, phagocytic, agranular leukocyte.

772. neutrophil

A granular leukocyte that fails to stain with either acidic or basic stains.

773. oxyhemoglobin

Hemoglobin to which oxygen is bound.

774. pernicious anemia

Anemia due to a lack of intrinsic factor and therefore vitamin B12.

775. plasma

The fluid part of blood including inactive clotting factors.

776. plasmin

An enzyme built into blood clots as they form that gradually dissolves them.

777. plasminogen

Inactive plasmin.

778. platelet

A megakaryocyte fragment in blood that participates in blood clotting reactions.

779. prothrombin

Inactive thrombin.

780. serum

The fluid part of blood after removal of formed elements and clotting factors.

781. sickle cell anemia

An inherited anemia in which erythrocytes sickle under low oxygen conditions.

782. spectrin

A protein that maintains flexibility of erythrocyte membranes.

783. streptokinase

An enzyme that digests blood clots used to treat coronary occlusion.

784. thrombin

An enzyme that activates fibrinogen to fibrin in the blood clotting mechanism.

785. thrombocyte

A platelet.

786. thrombus

A blood clot that is stationary in a blood vessel wall.

787. tissue plasminogen activator (tPA)

A substance secreted by many tissues that activates plasminogen to plasmin.

788. tissue thromboplastin

A substance from injured tissue that initiates extrinsic blood clotting.

789. transferrin

An iron-transport protein in plasma.

Fluid Balance / Waste Removal

790. acid-base balance

Maintenance of body fluid pH withing a normal range.

791. acidosis

Too low pH of body fluids because of accumulation of acid or carbon dioxide.

792. ammonotelic

Pertaining to excretion of ammonia as the endproduct of nitrogen metabolism.

793. archinephric duct

Primitive kidney duct.

794. Bowman's capsule

Glomerular capsule of kidney.

795. brush border

Presence of microvilli along cell membrane that increase surface.

796. clearance

Rate at which the kidneys can remove a substance from the blood.

797. collecting duct

One of many ducts that receive filtrate from kidney tubules.

798. countercurrent

A mechanism in which fluids flowing out of the system affect those flowing in.

799. electrolyte

Substance that ionizes and conducts electricity.

800. excretion

Elimination of a waste product.

801. fenestrated

Having one or more openings; having windows.

802. fluid regulation

Maintenance of body fluid volumes within normal ranges.

803. glomerulus

A capillary tuft surrounded by a glomerular capsule.

804. loop of Henle

U-shaped segment of a nephron where sodium chloride becomes concentrated in peritubular fluid.

805. malpighian tubule

Excretory organ for nitrogenous wastes of an arthropod.

806. mesonephros

Temporary embryonic kidney.

807. metanephridium

An advanced kind of invertebrate excretory structure.

808. metanephros

The embryonic kidney from which the mammalian functional kidney is derived.

809. nephridium

An excretory organ found in many invertebrates, especially segmented worms.

810. nephron

The functional unit of a mammalian kidney.

811. nephrostome

Ciliated opening from nephridium.

812. net filtration pressure

Pressure pushing materials out of a blood vessel.

813. opisthonephros

Adult kidney of most fishes and amphibians.

814. osmoregulation

Maintenance of internal fluid homeostasis by active transport.

815. papilla

A nipple-shaped projection.

816. peritubular

Surrounding a tubule, as in the kidney.

817. pronephros

The most primitive embyronic kidney among vertebrates.

818. protonephridium

Flame-cell excretory organ of many invertebrates.

819. renal

Of the kidney.

820. renin

Kidney secretion that activates angiotensinogen to angiotensin I.

821. renin-angiotensin mechanism

A mechanism that increases blood pressure and blood volume when either falls below normal.

822. thirst

Desire for water or other fluid.

823. urea

Nitrogenous waste from protein metabolism.

824. ureotelic excretion

Excreting nitrogen mainly as urea.

825. ureter

Tube through which urine flows from a kidney to the urinary bladder.

826. urethra

Tube through which urine flows from the urinary bladder outside the body.

827. uric acid

Nitrogenous waste from nucleic acid metabolism.

828. uricotelic

Concerning excretion of nitrogen mainly as uric acid.

829. urinary bladder

Distensible sac where urine is stored.

830. urine

Excretory product containing metabolic wastes.

831. water balance

A state in which water intake and water output are equal.

832. water expulsion vesicle

Organelle that osmoregulates fresh water protozoa.

Neurons / Neural Functions

833. **acetylcholine**

A neurotransmitter released by many axons, especially those that control skeletal muscles.

834. **action potential**

Wave of change in electrical potential across the cell membrane of an excited cell; impulse.

835. **adrenergic**

Concerning a neuron that releases norepinephrine (adrenalin).

836. **afferent**

Leading toward.

837. **association neuron**

A neuron that relays impulses from sensory to motor neurons, especially in the spinal cord.

838. **autonomic nervous system**

A nervous system component that regulates internal organ functions and involuntary processes.

839. **axon**

The part of a neuron that typically carries impulses away from the cell body toward another neuron.

840. **axon terminal**

The end of an axon from which neurotransmitter is released.

841. **catecholamine**

A class of amines that act as chemical messengers; dopamine, epinephrine, and norepinephrine.

842. **central nervous system (CNS)**

Brain and spinal cord.

843. cephalization

The concentration of sensory organs and nervous tissue at the head end of an animal.

844. cholinergic

Relating to a neuron whose terminals release acetylcholine.

845. cholinesterase

An enzyme that degrades acetylcholine.

846. cholinesterase inhibitor

A substance that blocks cholinesterase action.

847. craniosacral

Of the cranial and sacral regions.

848. dendrite

A cytoplasmic process of a neuron that commonly receives signals from other neurons.

849. dura mater

The tough outermost meninges that surrounds the brain, spinal cord, and other meninges.

850. efferent

Leading away from.

851. facilitation

Promotion of any process.

852. ganglion

An aggregation of cell bodies in the peripheral nervous system.

853. innervation

Nerve supply.

854. interneuron

A neuron that relays impulses between a sensory and a motor neuron, typically found in the spinal cord.

855. isthmus

Constriction; neck.

856. membrane potential

An electrical potential (potential difference between) the inside and outside of a membrane.

857. mesencephalon

A region of the embryonic brain near the middle of the brain.

858. metencephalon

Embryonic brain region from which the cerebellum and pons arise.

859. mixed nerve

A nerve with both sensory and motor fibers.

860. monosynaptic

Having a single junction between neurons.

861. motor neuron

A neuron that carries impulses toward a muscle or gland.

862. myelin

An insulating substance deposited around axons.

863. nerve

A bundle of axons covered with a connective tissue sheaths.

864. nerve impulse

Self-propagation wave of depolarization in a neuron.

865.   nerve net

Diffuse network of nerve cells.

866.   neural oscillator

A process that controls movements or other behaviors.

867.   neurilemma

The Schwann cell membrane.

868.   neurite

General name for nerve fiber.

869.   neuroglial cell

Supporting cell of the nervous system.

870.   neuron

Nerve cell that relays action potentials.

871.   neurosecretion

Chemical substance produced by a neuron.

872.   neurotransmitter

A chemical substance from one neuron that transmits a signal to another neuron at a synapse.

873.   Nissl granules

Granules containing RNA and endoplasmic reticulum in nerve cell bodies and dendrites.

874.   node of Ranvier

A gap in an axon's myelin sheath.

875.   norepinephrine

A neurotransmitter of the sympathetic division of the autonomic nervous system and of some brain neurons.

876. pattern generator

A control mechanism for repetitive movements.

877. perikaryon

Substance around a nucleus; the cell body of a neuron.

878. plexus

A network of nerves or blood vessels.

879. postganglionic

Concerning a neuron that receives signals across a synapse in a ganglion; second neuron in an autonomic pathway.

880. postsynaptic

Referring to a neuron that receives a neurotransmitter at a synapse.

881. preganglionic

Referring to an neuron that sends a signal through a ganglion.

882. presynaptic

Referring to a neuron that releases a neurotransmitter at a synapse.

883. principle of forward conduction

A rule that signals travel along axons toward the next neuron in a pathway.

884. reflex

An automatic involuntary response to a stimulus.

885. refractory period

Period of time during which a previously stimulated neuron cannot respond to another stimulus.

886. response

An action elicited by a stimulus.

887. resting potential

Electrical potential across the membrane of an unstimulated cell.

888. root

A base or foundation.

889. saltatory

Leaping.

890. Schwann cell

Neuroglial cell that forms the myelin sheath on peripheral nerve fibers.

891. sensory

Concerned with sensation.

892. serotonin

A substance secreted as a brain neurotransmitter and a gut hormone.

893. soma

Cell body.

894. somatic

Of the body.

895. stimulus

An event that typically elicits a response.

896. stretch reflex

Muscle contraction following stimulation of stretch receptors in a muscle or its tendon.

897. sympathetic chain ganglion

An aggregation of the cell bodies of postsynaptic neurons of the sympathetic division.

898. sympathetic division

The part of the autonomic nervous system that can respond to stressful situations.

899. synapse

A junction where a signal passes from one neuron to the next in a pathway, usually by neurotransmitter diffusion.

900. synaptic

Of a synapse.

901. thalamus

Subcortical gray matter near the anterior end of the brain stem.

902. threshold

Level of stimulus required to produce a response.

903. Wallerian degeneration

Axon disintegration in an injured neuron distal to the point of injury.

904. white matter

Myelinated nerve fibers of the central nervous system.

905. withdrawal reflex

A reflex that leads to flexion and removal of a limb from a painful stimulus.

Central Nervous System

906. Alzheimer's disease

A degenerative neurological disorder associated with memory loss and behavioral changes.

907. arachnoid

Spiderlike, as the delicate middle meninges that covers the brain and spinal cord.

908. arbor vitae

Treelike pattern of cerebellar white matter.

909. astrocyte

A star-shaped neuroglial cell in the central nervous system.

910. biofeedback

Use of signals indicating levels of autonomic process to control the process.

911. brain

Enlarged anterior end of the central nervous system.

912. brain stem

Brain parts that relay impulses to and from the cerebrum, cerebellum, and other brain structures.

913. brain wave

An electrical signal detected on the scalp that represents brain activity.

914. Broca's motor speech area

A functional area of the cerebrum where thoughts are translated into speech.

915. cerebellum

A brain component behind the cerebrum and above the pons concerned with the coordination of movements.

916. cerebrospinal fluid

A clear fluid in spaces within and around the central nervous system.

917. cerebrovascular

Relating to blood vessels of the brain.

918. cerebrum

The largest brain component; responds to sensory impulses and carries out mental processes.

919. circle of Willis

A ring of blood vessels at the base of the brain by which blood reaches alternate circulatory pathways in the brain.

920. consciousness

Awareness of signals from the sense organs.

921. corpus callosum

An aggregation of myelinated neural fibers that carry impulses between the two cerebral hemispheres.

922. corpus striatum

Subcortical mass of neuron cell bodies and fibers in the base of each cerebral hemisphere.

923. cortex

Outer portion of an organ; literally, bark.

924. diencephalon

Brain subdivision between telencephalon and mesencephalon that contains thalamus, epithalamus, and hypothalamus.

925. gray matter

Unmyelinated tissue (mainly cell bodies) in the central nervous system.

926. gyrus

A fold, such as a convolution of the cerebrum.

927. hippocampus

A part of the limbic system in the temporal region and concerned with emotion and memory.

928. hypothalamus

A part of the brain that connects and serves both the nervous and endocrine systems.

929. lateralization

Difference in function of the two sides of a bilateral structure, especially the cerebrum.

930. limbic system

A part of the brain mainly associated with emotions.

931. medulla oblongata

A part of the brain continuous with the spinal cord.

932. meninges

Membranes that surround the brain and spinal cord.

933. microglial cell

Small supporting cell of the central nervous system.

934. myelencephalon

Embryonic brain region from which the medulla oblongata arises.

935. neocortex

The most recently evolved part of the cerebrum.

936. neurofibrillary tangles

Masses of disorderly neural fibers in the brains of Alzheimer patients.

937. Parkinson's disease

Muscle rigidity and tremors due to a dopamine deficiency in the brain.

938. perception

Conscious interpretation of information from sensory receptors.

939. pia mater

A delicate membrane on the surface of the brain and spinal cord.

940. pineal gland

A gland between the cerebral hemispheres associated with regulating circadian rhythms.

941. pons

A part of the brain stem associated with the cerebellum.

942. Purkinje cell

A cerebellar neuron that synapses with a very large number of other cells.

943. rapid-eye-movement (REM) sleep

Sleep interval in which the eyeballs move and the EEG resembles wakefulness; paradoxical sleep.

944. reticular activating system (RAS)

A brain stem structure involved in maintaining consciousness.

945. sensation

An impression conveyed.

946. subconscious

Partially conscious; below the level of consciousness.

947. telencephalon

Developmental brain region from which cerebrum and olfactory bulbs arise.

948. visual area

A region of the occipital lobe of the cerebrum that receives and processes signals from the retina.

Sensory Organs

949. accommodation

Adjustment of the focal distance of the eyes to see close objects clearly.

950. adaptation

Decrease in excitability of sensory receptors after a period of continuous, constant intensity stimulation.

951. amplitude

Signal strength; the intensity of a sound.

952. ampulla

A dilation in a passageway.

953. anesthetic

Agent that produces temporary loss of sensation.

954. aqueous humor

Watery substance in the anterior cavity of the eye.

955. astigmatism

Blurred vision due to irregular curvature of one or more refractive surfaces in the eye.

956. auditory

Of hearing.

957. basilar membrane

Inner ear membrane associated with sound receptors.

958. binocular vision

Sight using two eyes that allows perception of three-dimensionality.

959. blind spot

Region of the retina that lacks light receptors where optic nerve fibers leave the eye.

960. cataract

Opacity of the eye lens.

961. cerumen

Earwax.

962. chemoreceptor

A receptor that responds to certain chemical substances.

963. choroid

Vascular middle layer of the eyeball.

964. ciliary body

Anteriormost part of the choroid layer, which contains ciliary muscles that participate in accommodation.

965. cochlea

Snail-shaped bony part of the inner ear.

966. conduction deafness

Loss of hearing due to impaired transmission of vibrations to sound receptors.

967. cone

A light receptor that responds to a certain color.

968. conjunctiva

A mucous membrane lining the eyelids and covering the anterior eyeball surface.

969. convergence

Coming together.

970. cornea

A transparent part of the anterior eye surface.

971. cupula

A domelike cup-shaped structure.

972. decibel

A unit on the logarithmic scale of sound intensity.

973. endolymph

Watery fluid within the membranous labyrinth of the inner ear.

974. exteroceptor

A sensory receptor that detects environmental changes.

975. fovea centralis

A pit in the retina that contains only cone receptors; region of greatest visual acuity.

976. frequency

Number of occurrences of an event in a give period, such as vibrations per second of sound waves.

977. glaucoma

A disorder in which aqueous humor accumulates and exerts excessive intraocular pressure.

978. gustatory

Pertaining to the sense of taste.

979. habituation

Gradual adaptation to a continuing stimulus.

980. hair cell

A kind of sensory receptor with an easily stimulated thin projection.

981. iris

Muscular diaphragm anterior to the eye lens that regulates the amount of light entering the eye.

982. kinesthetic

Concerned with sensing movement.

983. labyrinth

A maze; part of the inner ear.

984. lacrimal

Of tears.

985. lateral inhibition

Suppression of hair cells adjacent to those stimulated, which makes signals from stimulated receptors clearer.

986. lateral line organ

Sensory receptors found on the body surface of certain fish that allows them to detect objects in the water.

987. law of adequate stimulus

A law stating that a receptor responds only if it receives a sufficiently strong stimulus.

988. lens

Transparent, biconcave structure behind the iris that focuses eye on far or near objects by changing shape.

989. lysozyme

An enzyme in tears that can destroy microbes.

990. macula

An inner ear structure that contains sensory receptors for static equilibrium.

991. mechanoreceptor

A receptor that responds to mechanical pressure.

992. medulla

The inner core of an organ.

993. Meniere's disease

A semicircular canal inflammation that leads to sight, hearing, and balance disorders.

994. modiolus

A structure that supports the cochlea.

995. neuromast

Cluster of sensory hair cells covered by a gelatinous cupula.

996. nociceptor

A receptor that responds specifically to painful stimuli.

997. olfactory

Of the sense of smell.

998. olfactory epithelium

Cell layer in nasal cavity that contains smell-sensing cells.

999. opsin

A protein that combines with retinine in the retina.

1000. optic chiasma

A site anterior to the pituitary where medial fibers of each optic nerve cross from one side of the body to the other.

1001. optic disk

Region of the retina lacking receptors where optic nerve fibers leave the eyeball; blind spot.

1002. organ of Corti

The inner ear structure where sound receptors are located.

1003. osphradium

Sensory organ in the mantle cavity of some mollusks.

1004. ossicle

Sound-transmitting bone or bone-like structure.

1005. otolith

A small calcium carbonate particle in a receptor for static equilibrium; an ear stone.

1006. oval window

A membrane-covered opening from the middle to the inner ear across which the stapes transmits vibrations.

1007. Pacinian corpuscle

A receptor that responds to pressure.

1008. perilymph

Fluid found between membranous labyrinth and bony covering of the inner ear.

1009. photon

The smallest unit of light energy.

1010. photoreceptor

A receptor that responds to light.

1011. pinna

The flap-like part of the ear projecting from the head.

1012. pitch

A sound quality determined by vibration frequency.

1013. proprioceptor

Any sensory receptor in a muscle, joint, or tendon that detects position or movement.

1014. rarefaction

Decreased density.

1015. refraction

Bending of light rays as they pass from a medium of one density to a medium of a different density.

1016. retina

The innermost layer of the eye, which contains light receptors.

1017. retinene

A carotenoid pigment that binds to opsin.

1018. rhodopsin

A light-sensitive protein found in rods of the retina.

1019. rod

A receptor in the retina that responds to different intensities of light but not to color.

1020. round window

A membrane-covered opening between the middle and inner ear.

1021. sclera

Tough fibrous outer wall of vertebrate eyeball.

1022. semicircular canal

One of three pairs of fluid-filled inner ear passageways that detect head movements.

1023. static equilibrium

Maintenance of balance while the head is stationary.

1024. statocyst

An organ of balance in a crayfish.

1025. stigma

Eyespot.

1026. taste bud

A structure on the tongue containing taste receptor cells.

1027. tear

A fluid released from tear glands.

1028. tectorial membrane

Membrane in roof of the organ of Corti.

1029. thermoreceptor

A receptor that detects temperature changes.

1030. transducin

Enzyme involved in visual process.

1031. transduction

The conversion of a signal from one type to another, such as from chemical to electrical.

1032. tympanic membrane

Eardrum; membrane between the external and middle ear.

1033. tympanum

Sound receptor found in terrestrial animals.

1034. utricle

Large chamber in the ear vestibule that contains receptors for equilibrium.

1035. vestibule

A space or cavity near the entrance of a canal.

1036. visceroceptor

A receptor in or near an internal organ.

1037. vitreous humor

Glassy substance in the posterior cavity of the eye.

Body Defenses

1038. acquired immune deficiency syndrome (AIDS)

A viral disease that severely impairs immunity.

1039. acquired immunity

Disease resistance obtained from another's antibodies.

1040. active immunity

Disease resistance obtained by the immune system responding to a microorganism or a vaccine.

1041. agglutination

Clumping of cells such as erythrocytes in an immune reaction.

1042. agglutinin

An antibody in an agglutination reaction.

1043. agglutinogen

An antigen that elicits an agglutination reaction.

1044. allergen

A substance capable of eliciting an allergic reaction.

1045. allergy

Unusual sensitivity to a normally harmless concentration of a substance.

1046. anaphylaxis

A severe allergic reaction to a substance to which one has been previously sensitized.

1047. antibody

A protein released in response to an antigen that can inactivate the antigen.

1048. antigen

A substance that elicits a response from the immune system.

1049. antiserum

Fluid part of blood containing antibodies for a particular antigen.

1050. antitoxin

Antibody that neutralizes a toxin.

1051. artificially acquired active immunity

Disease resistance obtained by stimulating the immune system with a vaccine.

1052. artificially acquired passive immunity

Temporary disease resistance obtained by receiving another's antibodies.

1053. B lymphocyte

A lymphocyte that produces plasma cells, which in turn produce antibodies.

1054. bacterial antagonism

Inhibition of growth of one bacterial species by another species.

1055. bursa of Fabricius

A structure found in birds where B lymphocyte differentiation was first identified.

1056. cell-mediated immunity

Disease resistance involving direct destruction of antigenic cells.

1057. chemotaxis

The act of chemical stimuli to attract or repel; a process that causes some leukocytes to migrate to sites of injury.

1058. clonal selection theory

Explanation of how lymphocytes are sensitized to a certain antigen and how immune tolerance for self arises.

1059. clone

A group of genetically like cells derived from a single parent cell.

1060. coccyx

The caudal end of the spinal column; tail bone.

1061. complement

A group of plasma enzymes that catalyze a sequence of reactions against many different kinds of foreign matter.

1062. heterogeneity

Diversity.

1063. heterograft

Moving tissue from one species to another.

1064. histamine

A derivative of the amino acid histidine released by injured cells that causes vasodilation and bronchial constriction.

1065. homograft

Moving tissue from one organism to another of the same species.

1066. humoral immunity

Disease resistance produced by antibodies.

1067. hybridoma

A cell made by fusing parts of two cells.

1068. hypersensitivity

Abnormal reaction to a substance, as occurs in allergy and certain other immune reactions.

1069. IgA

An immunoglobulin in secretions.

1070. IgD

An immunoglobulin of unknown function.

1071. IgE

An immunoglobulin responsible for allergic responses.

1072. IgG

An immunoglobulin in blood and primarily responsible for resisting infection.

1073. IgM

A multiunit immunoglobulin most abundant early in an immune response.

1074. immune

Disease resistant.

1075. immunity

A state of disease resistance.

1076. immunization

Use of a vaccine or other procedure to create immunity.

1077. immunodeficiency

An absence or lack of a normal immune function.

1078. immunoglobulin

A protein that can bind with a foreign substance; antibody that binds with an antigen.

1079. immunologic tolerance

Ability of organism to accept an antigen without reacting.

1080. immunology

The study of immunity and immune reactions.

1081. immunosuppression

A procedure used to lessen an immune response.

1082. immunotoxin

An antibody bound to a toxic drug.

1083. inflammation

Localized response to tissue injury, usually involving redness, swelling, increased temperature, and pain.

1084. interferon

A protein released by virally-infected cells that causes adjacent cells to make an antiviral protein.

1085. interleukin

A substance that facilitates or enhances an immune reaction.

1086. kinin

A substance that stimulates events in the inflammatory process.

1087. lymphokine

A substance that stimulates activity of lymphocytes.

1088. macrophage

A large phagocytic cell in connective tissue.

1089. melanin

A dark brown pigment of hair and skin.

1090. naturally acquired active immunity

Immunity produced by having a disease.

1091. naturally acquired passive immunity

Immunity based on antibodies transferred across the placenta or in breast milk.

1092. passive immunity

Temporary disease resistance derived from antibodies from another organism (human or other).

1093. phagocytosis

Engulfment into a vacuole and digestion by a scavenger cell.

1094. plasma cell

An antibody-producing cell derived from a B lymphocyte.

1095. pus

A product of inflammation consisting of debris from dead leukocytes and microorganisms.

1096. reticuloendothelial system

System of phagocytic cells in vertebrates.

1097. Rh factor

Agglutinogen (as from fetus) that elicits antibody production (by mother) that can damage future fetus.

1098. scab

A crust over a superficial wound.

1099. scar

Connective tissue that has replaced an injured tissue unable to replace itself.

1100. secondary response

Rapid release of antibodies because of previous sensitization to an antigen.

1101. sensitization

Rendering a lymphocyte sensitive to a foreign substance.

1102. T lymphocyte

A thymus-processed lymphocyte that can differentiate into several kinds of T cells.

1103. theory of immune surveillance

A possible way the body finds malignant cells and destroys them.

1104. toxin

Poisonous substance.

1105. vaccine

Commercial product that induces the body to develop immunity to a particular disease.

Endocrine Functions

1106. adenohypophysis

Anterior pituitary gland.

1107. adrenal

Above the kidney; a gland lying superior to the kidney.

1108. adrenalin

A hormone secreted by the adrenal glands.

1109. adrenocorticotropic hormone

A hormone that stimulates the adrenal cortex to secrete hormones.

1110. aldosterone

An adrenocortical hormone that increases reabsorption of sodium.

1111. anabolic steroid

A synthetic hormone that increases muscle size.

1112. androgen

A molecule with male hormone activity.

1113. antidiuretic hormone (ADH)

A hypothalamic hormone stored in the posterior pituitary gland that stimulates water conservation by the kidneys.

1114. calcitonin

A hormone that lowers blood calcium.

1115. corpus allata

Endocrine gland of insects that secretes juvenile hormone.

1116. corpus cardica

Aggregate of nerve cells in insects that releases brain hormones.

1117. cortisol

An adrenocortical hormone that helps regulate carbohydrate metabolism and counteracts inflammation.

1118. endocrine

Concerning a ductless gland.

1119. epinephrine

Main hormone from the adrenal medulla.

1120. follicle-stimulating hormone (FSH)

A hormone that stimulates maturation of ova and sperm.

1121. gastrin

A hormone from the stomach lining that circulates in the blood and stimulates HCl secretion.

1122. glucagon

A hormone that raises blood glucose.

1123. glucocorticoid

A hormone that helps to regulate carbohydrate metabolism.

1124. goiter

Enlargement of thyroid gland from overactivity or lack of iodine.

1125. hormone

A regulatory substance from an endocrine cell that is transported in the blood to its target cells.

1126. hypophysis

The pituitary gland.

1127. insulin

A hormone from the pancreas that causes cells to take in glucose and stimulates protein synthesis.

1128. islet of Langerhans

Cluster of hormone-secreting cells in the pancreas.

1129. luteinizing hormone (LH)

A hormone that helps to cause ovulation and other reproductive processes.

1130. mineralocorticoid

A hormone that regulates mineral metabolism.

1131. negative feedback

A control system in which a system's output inhibits activity of the system.

1132. neurohypophysis

The posterior, neural part of the pituitary gland.

1133. noradrenalin

Norepinephrine.

1134. oxytocin

A hormone from the hypothalamus that stimulates uterine contractions and milk let down.

1135. parasympathetic division

Autonomic component that accelerates digestion and other functions not essential to a response to stress.

1136. parathormone

A hormone from the parathyorid gland that decreases blood calcium.

1137. parathyroid glands

Glands imbedded in the thyorid gland.

1138. pituitary

Gland just below the hypothalamus that produces a variety of hormones.

1139. positive feedback

Output from a system that accelerates change in the system.

1140. progesterone

A hormone that helps to maintain pregnancy.

1141. prolactin

A hormone that stimulates milk secretion.

1142. prostaglandin

A substance derived from the fatty acid arachidonic acid that acts over short distances as a chemical messenger.

1143. puberty

A period during which sexual maturity is achieved.

1144. Rathke's pouch

An embryonic structure from which the anterior pituitary develops.

1145. relaxin

A hormone from the corpus luteum of pregnancy.

1146. releasing hormone

Secretion from hypothalamus that regulates release of anterior pituitary hormone.

1147. second messenger

Molecule in a cell that responds to a hormone and produces a specific effect on a target cell.

1148. somatostatin

Growth-hormone inhibiting hormone.

1149. somatotropin

Growth hormone.

1150. stress

A condition produced by a variety of injurious agents that affects many body systems.

1151. stressor

An agent or event that produces stress.

1152. target cell

A cell that can respond to a certain hormone.

1153. testosterone

A male hormone.

1154. thymosin

A hormone secreted by the thymus gland.

1155. thymus gland

A gland that processes and activates T lymphocytes before it regresses during puberty.

1156. thyroid gland

A gland in the throat that produces metabolism regulating hormones.

1157. thyroid-stimulating hormone (TSH)

A hormone that stimulates hormone secretion by the thyroid gland.

1158. thyroxine

Hormone produced by the thyroid gland that regulates metabolic rate.

1159. tract

A bundle of myelinated neurons in the brain or spinal cord.

1160. tropic

Influencing another organ or process.

Reproductive Organs

1161. amenorrhea

Absence of menstruation.

1162. amniocentesis

Procedure for taking a sample of amniotic fluid from around a fetus to detect genetic or developmental defects.

1163. areola

Pigmented region around the nipple in a mammary gland.

1164. Bartholin's gland

Vulvovaginal gland.

1165. cervix

Neck; narrow end of the uterus adjacent to the vagina.

1166. clitoris

Small erectile organ homologous to the penis located in the anterior vulva.

1167. colostrum

First fluid from a mammary gland after childbirth.

1168. corpus albicans

A scar on an ovary that remains after the degeneration of a corpus luteum; literally, white body.

1169. corpus luteum

Follicular cells that produce hormones after ovulation; literally, yellow body.

1170. Fallopian tube

Uterine tube.

1171. fallopian tube

Uterine tube.

1172. genitalia

Genital or sex organs.

1173. glans

The cone-shaped tip of the penis or clitoris.

1174. gonorrhea

A sexually transmitted disease caused by Neisseria gonorrhoeae.

1175. graafian follicle

An ovarian follicle.

1176. labia

Lip-shaped structures.

1177. lactation

Synthesis and secretion of milk.

1178. lactiferous

Making or conveying milk.

1179. mammary gland

Gland that synthesizes and secretes milk.

1180. menarche

Onset of menstruation during puberty.

1181. menopause

Cessation of menstruation.

1182. menstrual cycle

Repetitive sequence of events involving ovulation and preparation of the uterus for implantation.

1183. menstruation

Periodic discharge of blood, tissue debris, and fluid from the uterus.

1184. mons pubis

Fatty area covered with pubic hair.

1185. ovary

A female gonad.

1186. oviduct

Uterine tube or Fallopian tube.

1187. ovulation

Sudden expulsion of an ovum from a follicle.

1188. parturition

Childbirth.

1189. placenta

Structure attached to the uterine wall that provides nutrients and removes wastes for a developing fetus.

1190. postovulatory

After ovulation.

1191. preovulatory

Before ovulation.

1192. prepuce

Foreskin of the penis.

1193. proliferative phase

A part of the menstrual cycle in which endometriual cells divide and increase in number.

1194. reproductive engineering

Human-designed procedures used to alter the reproductive process.

1195. scrotum

A pouch in which the testes are located.

1196. semen

A fluid containing sperm and secretions from the male reproductive glands.

1197. seminal receptacle

Female reproductive structure in which sperm are stored after mating.

1198. seminal vesicle

A convoluted saclike organ near the ductus deferens that adds secretions to semem.

1199. seminiferous tubule

A coiled tubule within a testis in which sperm are produced.

1200. Sertoli's cell

Cell type in seminiferous tubules that nourishes developing sperm.

1201. spermatic cord

A cord extending from the scrotum to the inguinal ligament that includes ductus deferens, blood vessels, and nerves.

1202. syphilis

A sexually transmitted disease caused by the spirochete Treponema pallidum.

1203. theca

Sheath, such as that covering an ovarian follicle.

1204. umbilicus

Site where umbilical cord was attached to fetus; navel.

1205. uterine

Of the uterus.

1206. uterine tube

A tube between an ovary and the uterus.

1207. uterus

A hollow, pearshaped organ where a fetus develops.

1208. vagina

Passageway from the uterus.

1209. vas deferens

A duct between the epididymis and the ejaculatory duct; ductus deferens.

1210. Wharton's jelly

Soft, pulpy connective tissue of the umbilical cord matrix.

Meiosis / Fertilization

1211. acrosome

Dense anterior part of a sperm that contains enzymes needed to penetrate an ovum.

1212. alternation of generations

A reproductive cycle involving gametophyte and sporophyte generations of plants.

1213. artificial insemination

Method for introducing sperm into the female reproductive tract without sexual intercourse.

1214. asexual reproduction

Reproduction not involving the union of gametes.

1215. autogamy

Fusion of two haploid gametes within the organism that produced them.

1216. binary fission

Reproduction in which an individual divides into two equal parts.

1217. bisexual

Having both male and female sex organs; attracted to both sexes.

1218. budding

Asexual reproduction in which a new organism or cell pinches off from the parent.

1219. conjugation

Mating between two simple organisms (ciliates) with exchange of micronuclei.

1220. contraception

Prevention of the union of sperm and egg.

1221. copulation

Physical joining of animals while sperm are transferred from one to the other.

1222. dioecious

Having separate male and female individuals in a species.

1223. diploid

Having paired chromosomes.

1224. ejaculation

Ejection of semen.

1225. epididymis

Coiled duct system in testes where sperm are stored.

1226. erection

Rigid state of the penis.

1227. estrogen

A kind of female hormone that stimulates development of sex organs and secondary sexual characteristics.

1228. estrus

Period of sexual receptivity in many female mammals.

1229. external fertilization

The union of sperm and eggs outside parental bodies.

1230. female pronucleus

Nucleus of an ovum that fuses with the nucleus of a sperm.

1231. fertilization

The union of egg and sperm.

1232. fertilization membrane

The vitelline membrane after fertilization.

1233. fertilized ovum

An ovum already penetrated by a sperm.

1234. fission

Asexual reproduction in which a cell divides into two.

1235. follicle

Small sac as that which contains the egg in a mammalian ovary.

1236. fragmentation

Breaking of an animal into pieces in a kind of asexual reproduction.

1237. fusion nucleus

Diploid nucleus produced by fusion of egg and sperm nuclei.

1238. gamete

Haploid cell; an ovum or sperm.

1239. gametogenesis

The process of forming gametes.

1240. gamont

Cell that exchanges nuclei with another of the same species prior to reproduction.

1241. gemmule

Asexual reproductive body with protective cover in a sponge.

1242. germinal epithelium

Epithelial cells that divide to form gamete-producing cells.

1243. gonad

Organ that produces gametes.

1244. gonopore

External opening from a reproductive system.

1245. hermaphroditic animal

An animal having both male and female sex organs.

1246. heterogamy

Reproduction involving gametes that differ in size and shape.

1247. implantation

Attachment of an embryo to the endometrium of the uterus.

1248. internal fertilization

The union of egg and sperm within a parental, usually female, body.

1249. isogamy

Having structurally identical male and female gametes.

1250. luteum

Yellow.

1251. male pronucleus

Nucleus of a sperm that has penetrated an ovum.

1252. meiosis

Cell division that gives rise to haploid cells.

1253. mictic

Concerning a haploid egg that can be fertilized or the female that produces it.

1254. monoecious

Concerning species in which individuals have both male and female organs.

1255. nondisjunction

Failure of replicated chromosomes to separate.

1256. oocyte

Cell that gives rise to an ovum.

1257. oogenesis

Process of producing an ovum.

1258. oogonium

Female germ cell.

1259. orgasm

An intense pleasurable culmination of sexual intercourse.

1260. oviparous

Pertaining to an animal that produces eggs that develop outside the mother's body.

1261. ovoviviparous

Pertaining to an animal that produces eggs that develop inside the mother's body.

1262. ovum

Female gamete.

1263. paedogenesis

Parthogenesis or other reproduction by larval organisms.

1264. penis

Male copulatory organ.

1265. polar body

A nonfunctional cell from unequal meiotic divisions during animal oogenesis.

1266. primary follicle

An early, immature stage of an ovarian follicle.

1267. primordial

Original or primitive.

1268. prostate

Largest accessory gland in the male reproductive system of mammals.

1269. protandry

Sexual maturity in males before that in females.

1270. protogyny

Sexual maturity in females before that in males.

1271. reproductive efficiency

The ratio of survivors to the total number of eggs produced.

1272. schizogony

Asexual reproduction in which a multinucleate cell divides into many uninucleate cells.

1273. sexual dimorphism

The presence of structural differences in males and females.

1274. sexual reproduction

Reproduction involving the production and union of gametes.

1275. sperm

A male gamete.

1276. spermatid

An immature spermatozoan.

1277. spermatocyte

A cell from which sperm develop.

1278. spermatogenesis

The process of sperm formation.

1279. spermatogonium

Undifferentiated male germ cells from which sperm-producing cells arise.

1280. spermatophore

Package of sperm secreted from certain male reproductive tracts.

1281. spermatozoan

Male gamete, sperm.

1282. spermiogenesis

Maturation of spermatids to become sperm.

1283. spore

Resistant reproductive structure of some protozoa.

1284. sporogony

Asexual reproduction that produces spores or sporozoites.

1285. sterility

Inability to produce offspring.

1286. stroma

Fibrous connective tissue framework that supports an organ.

1287. synapsis

Alignment of homologous pairs of chromosomes.

1288. syngamy

Fusion of gametes in fertilization.

1289. testis

A male gonad.

1290. tetrad

Four copies of the same chromosome temporarily attached to each other.

1291. triploid

Having three sets of chromosomes.

1292. viviparous

Producing living young, as opposed to laying eggs.

1293. X organ

Organ in crustacea that produces regulatory hormones.

1294. zona pellucida

Translucent membrane around an oocyte in an ovarian follicle.

1295. zooid

Individual member of a colony of animals.

1296. zygote

Single cell resulting from a union of ovum and sperm; first cell of a new individual.

1297. zygotic meiosis

The occurrence of meiosis during the division of a zygote in sexually reproducing organisms.

Development

1298. adolescence

Period from the onset of puberty to adulthood.

1299. aging

Process of growing old.

1300. alecithal

Pertaining to an egg having no yolk.

1301. allantois

A fetal membrane that helps to form the umbilical cord.

1302. amictic

Concerning unfertilizable diploid eggs or female rotifers that produce them.

1303. amnion

Membrane around a fetus that fills with fluid and acts as a shock absorber.

1304. amniote

Organism whose embryos have an amnion and other embryonic membranes.

1305. amniotic

Of the amnion.

1306. amphiblastula

Larval sponge blastula containing two kinds of cells.

1307. anamniote

Animal whose embryos lack embryonic membranes.

1308. archenteron

A primitive gut cavity in an animal embryo.

1309. atresia

Blockage of or absence of a normal passage or cavity.

1310. blastocoele

Cavity in a blastula.

1311. blastocyst

Hollow ball of cells that arises early in embryonic development.

1312. blastoderm

Cell layer that encloses the blastocoel.

1313. blastodisc

Small cytoplasmic disc on an egg yolk that contains the egg.

1314. blastomere

A cell from early divisions during cleavage of a zygote.

1315. blastopore

Opening in the archenteron formed by gastrulation.

1316. blastula

A hollow ball of cells in the early embryonic development of an animal.

1317. cell lineage

Cell history through successive cleavage, which can be traced in mosaic embryos.

1318. centrolecithal

Arthropod egg having yolk in large sphere around central nucleus.

1319. chimera

Organism whose body contains cells with different genomes.

1320. chorion

Outermost fetal membrane, which is incorporated into the placenta.

1321. chorionic villi

Tufts of fetal blood vessels across which substances are exchanged with maternal blood.

1322. cleavage

Division into two equal parts; process by which a zygote develops into a multicellular ball.

1323. cleidoic egg

Self-sufficient egg of reptiles and birds that gives rise to minature adult without larval stage.

1324. congenital

Present at birth.

1325. delamination

Separation of ectoderm and endoderm during embyronic development.

1326. differentiation

The specialization of structures during embyronic development.

1327. diploblastic

Concerning an embryo that has two cell layers.

1328. dizygotic

Arising from two separate zygotes.

1329. ectoderm

The outermost germ layer in an embryo.

1330. embryo

Early developmental stage of an organism.

1331. embryonic disc

Trophoblast cells that give rise to the embryo.

1332. embryonic induction

Interaction of two embryonic tissues in which one affects the developmental potential of the other.

1333. endoderm

The innermost germ layer in an embryo.

1334. epiboly

Gastrulation in which smaller animal pole cells grow over and enclose cells of the vegetal pole.

1335. epigenesis

A theory that the development of an embryo consists of gradual elaboration and organization of parts.

1336. eutely

Having a nearly constant number of cells in the adult body.

1337. extraembryonic membrane

One of several membranes that surround a developing vertebrate embryo.

1338. fetus

Unborn child from two months development to birth.

1339. gastrula

The stage of a developing vertebrate embyro that follows the blastula.

1340. gastrulation

The developmental process of forming a gastrula.

1341. holoblastic cleavage

Undergoing complete cleavage of the zygote.

1342. induction

Influence exerted by one tissue over another during embryonic development.

1343. infancy

The period of time from 1 month to 2 years of age.

1344. ingression

Migration of outer cells into the blastocoel.

1345. inner cell mass

Cells that give rise to a mammalian embryo.

1346. involution

Inward migration and expansion of outer cells into area under other cells.

1347. isolecithal

Concerning an egg with yolk evenly distributed in cytoplasm.

1348. lanugo

Fine hair on the skin of a fetus.

1349. life cycle

Sequence of stages a species goes through from egg to egg or adult to adult.

1350. macromere

Larger cell produced by unequal cell division.

1351. meroblastic cleavage

Cleavage restricted to the cytoplasmic part of a yolk laden egg.

1352. mesoderm

The middle germ layer in an embryo.

1353. mesolecithal

Concerning an egg in which a moderate portion of yolk is concentrated in the vegetal pole.

1354. micromere

Smaller cell produced by unequal cell division.

1355. monozygotic

Arising from the same zygote.

1356. morphogenesis

Development of form, size and other features of a body or organ.

1357. morula

A solid ball of cells in early embryological development.

1358. neonate

A newborn infant.

1359. neoteny

Presence of larval-like features in adult; reproduction in larval form.

1360. neural crest

Cells that give rise to sensory neurons, adrenal medulla, and the autonomic nervous system.

1361. neural tube

A tube formed by invagination of ectoderm that gives rise to the nervous system.

1362. neurula

Embryonic stage in which the primitive nervous system forms.

1363. neurulation

Formation of a neural tube as a part of embryonic development.

1364. oligolecital

Pertaining to an egg with a small amount of yolk.

1365. organizer

A part of an embryo that influences the development of another part.

1366. organogenesis

The embryonic development of organs.

1367. parthenogenesis

The development of unfertilized eggs.

1368. planula

Free-swimming, ciliated larva having two germ layers.

1369. polyembryony

Formation of two or more embryos from one zygote.

1370. primitive streak

An area in an early embryo having a large amount of yolk that represents the blastopore in embryos with less yolk.

1371. radial cleavage

Developmental cell division in echinoderms and vertebrates with spindle axes parallel/perpendicular to polar axis.

1372. redia

Second larval stage of a fluke, which reproduces asexually.

1373. regeneration

The regrowth of a lost or injured part.

1374. segmentation

Division of body into a series of similar parts.

1375. senescence

Gradual loss of function with aging.

1376. somite

An embryonic segment of mesoderm.

1377. spiral cleavage

Developmental cell division in which cleavage planes are oblique to polar axis, as in annelids and mollusks.

1378. stereoblastula

Blastula in which blastocoel is filled with cells.

1379. stereogastrula

Gastrula in which the central cavity is filled with cells.

1380. subgerminal space

Shallow cavity under dividing cells in chicken egg.

1381. telolecithal

Having an abundant supply of yolk.

1382. trisomy

Condition of having three copies of a chromosome.

1383. trophoblast

Outer blastocyst layer that is connected with maternal tissue and gives rise to chorionic villi.

1384. vitellarium

Gland that synthesizes and secretes yolk.

1385. vitelline

Concerning yolk.

1386. vitelline membrane

A protective covering membrane of an egg.

1387. yolk

Food storage part of an egg.

1388. yolk sac

Embryonic bag containing nutrients.

Mendelian Genetics

1389. allele

One of several different genes for a trait that can occupy a particular site on a chromosome.

1390. allopolyploidy

Hybridization of two species to form a third species.

1391. aneuploidy

Having more or less than the normal diploid number of chromosomes.

1392. autosomal

Concerning paired (nonsex) chromosomes and the genetic information they carry.

1393. autosome

One of a pair of nonsex chromosomes.

1394. chromosomal abnormality

Detrimental change in the DNA configuration in a chromosome.

1395. codominance

The simultaneous expression of two alleles in one individual without blending.

1396. crossing-over

The exchange of corresponding segments of DNA during meiosis.

1397. differential migration

Alteration of gene frequencies in a population by population mobility.

1398. dihybrid

Involving two pairs of alleles.

1399. dominant allele

The allele expressed when it and a recessive allele are present.

1400. epistatic

One gene interfering with the effect of another gene.

1401. expressivity

Extent to which an inherited trait is manifested by an organism carrying genetic information for the trait.

1402. genetics

The study of heredity.

1403. genotype

Alleles of a single gene or all the genes carried by a particular individual.

1404. hemizygous

Having one of each kind of sex chromosome.

1405. heredity

Transmission of characteristics from one generation to the next.

1406. heterosis

Vigor produced in offspring of totally unrelated strains of an organism.

1407. heterozygous

Having unlike alleles for a trait.

1408. homologous

Having the same shape and structure; pairs of chromosomes having the same information.

1409. homozygous

Having like alleles for a trait.

1410. Huntington's chorea

A dominant hereditary disorder that causes degeneration of the nervous system.

1411. hybrid

An offspring produced by crossing populations differing in one or more traits.

1412. hybrid vigor

The display of increased fitness resulting from crossing of populations.

1413. incomplete dominance

The expression by blending of two alleles present at the same time in an organism.

1414. Klinefelter's syndrome

A condition due to the presence of XXY sex chromosomes.

1415. linkage

Degree to which genes are closely associated physically and thereby inherited together.

1416. linkage group

A group of genes physically close and inherited together.

1417. locus

Point on a chromosome where a particular gene is located.

1418. monohybrid

Pertaining to a genetic cross that differs in a single trait under study.

1419. mosaic

Organism with some cells containing different genetic information than other cells.

1420. multiple alleles

Three or more kinds of genetic information for a given trait.

1421. outbreeding

Mating of organisms of unrelated strains.

1422. penetrance

Frequency with which an inherited trait is expressed in organisms carrying the genetic information for it.

1423. phenocopy

Simulation of genetic traits of another genotype.

1424. phenotype

The appearance of an individual with respect to one or all inherited characteristics.

1425. pleiotropy

The influence of a single gene on more than one trait.

1426. ploidy

Pertaining to the number of sets of chromosomes in a cell.

1427. polygenic inheritance

A situation in which two or more genes, each with alleles, jointly affect the expression of a trait.

1428. polyploidy

Presence of more than two sets of chromosomes.

1429. principle of dominance

A Mendelian principle that one factor for a trait can mask or overpower another factor for the same trait.

1430. principle of unit characters

A Mendelian principle that individuals carry two factors for each trait.

1431. Punnett square

A table to illustrate the distribution of alleles in the offspring of heterozygtes.

1432. recessive

In genetics, a characteristic seen in the phenotype only when recessive allele is only one present in the genotype.

1433. selection

Favoring of certain alleles in a gene pool.

1434. sex-linked

Located on a sex chromosome.

1435. testcross

The mating of an organism with an unknown genotype to one with a homozygous recessive phenotype.

1436. tetraploidy

Having four sets of chromosomes.

1437. triploidy

Having three sets of chromosomes.

1438. Turner's syndrome

A condition due to having a single X chromosome (without another X or Y chromosome).

1439. variation

Divergence among individuals in a species.

Molecular Genetics

1440. anticodon

A three-base sequence of transfer RNA that fits with a particular codon on messenger RNA.

1441. antimetabolite

Substance that mimics and interferes with use of essential nutrient.

1442. auxotrophic

A mutant organism that has greater nutritional requirements than its normal counterpart.

1443. biotechnology

The use of a natural biological system to make a product or achieve a particular end.

1444. codon

A three-base sequence in messenger RNA derived from DNA and specifying amino acid placement in a protein.

1445. complementary base pairing

Bonding between certain bases in nucleic acid strands.

1446. cystic fibrosis

An inherited disorder in which thick mucus blocks respiratory and pancreatic passageways.

1447. deletion

Loss of one or more bases from a DNA strand.

1448. DNA ligase

An enzyme that attaches cut ends of DNA molecules.

1449. DNA polymerase

An enzyme that increases chain length in DNA synthesis.

1450. DNA replication

Synthesis of new DNA according to information in an existing DNA template.

1451. frameshift mutation

A DNA sequence change caused by adding or deleting bases.

1452. gene

Functional unit of heredity; a site on a chromosome that transmits a particular hereditary characteristic.

1453. gene amplification

Selective replication of certain genes.

1454. gene sequencing

Determination of the order of nucleotides in a gene.

1455. gene therapy

Biotechnical applications designed to treat genetic disease.

1456. genetic code

The three-base sequences in messenger RNA derived from a DNA template that determine amino acid order in proteins.

1457. genetic engineering

Use of human-designed procedures to alter genetic information.

1458. genetic equilibrium

A state of constancy in the frequency of alleles in a population.

1459. genetic load

The number of genetic defects present in a population.

1460. genetic screening

Search for genetic defects in fetuses, newborns, and prospective parents.

1461. genome

An organism's whole complement of DNA.

1462. genomic library

A collection of genetically engineered viruses carrying all the genes of a species.

1463. inducer

A regulatory molecule that promotes expression of a gene.

1464. inducible enzyme

An enzyme synthesized only in the presence of its substrate.

1465. insertion mutation

A change in DNA involving the addition of nucleotides at some point along the strand.

1466. inversion

A change in a chromosome resulting in a reordering of its genes.

1467. lagging strand

The DNA strand on which synthesis is discontinuous during replication.

1468. leading strand

The DNA strand that is replicated continuously.

1469. messenger RNA

A nucleic acid that carries information in the form of codons for the synthesis of a protein.

1470. missense mutation

A point mutation that replaces one amino acid for another in a protein.

1471. molecular hybridization

A procedure for determining similarity of nucleotide sequence between two nucleic acid molecules.

147

1472. mutagen

An agent that can alter DNA.

1473. mutation

A change in genetic information.

1474. nonsense mutation

A point mutation that produces a codon that stops protein synthesis, thereby creating a short nonfunctional peptide.

1475. Okazaki fragments

Short segments of single-stranded DNA synthesized on a lagging strand.

1476. oncogene

A gene that contributes to the development of cancer.

1477. operon

A group of bacterial genes that function together and are controlled by a regulator gene.

1478. point mutation

A change in a single base in a DNA molecule.

1479. purine

A nitrogenous base with two rings found in nucleic acids.

1480. pyrimidine

A nitrogenous base with one ring found in nucleic acids.

1481. recombinant DNA

DNA segments combined from two different organisms.

1482. regulator gene

A gene that directs the activity of a specific set of genes.

1483. replication

Duplication.

1484. repressor

A gene that prevents the action of a set of genes.

1485. restriction enzyme

An endonuclease that cuts double stranded DNA at sites having specific nucleotide sequences.

1486. restriction site

The site at which an endonuclease acts.

1487. reverse transcriptase

An enzyme that makes DNA according to an RNA template.

1488. ribosomal RNA (rRNA)

A nucleic acid that forms part of a ribosome.

1489. semiconservative replication

The replication of DNA in which each molecule consists of one new and one old strand.

1490. structural gene

A gene that produces a specific product.

1491. template

Pattern.

1492. transcription

The transfer of coded genetic information from DNA to mRNA.

1493. transfer RNA

RNA that carries amino acids and places them in specific sites in a growing peptide chain.

1494. translation

The process by which mRNA codons are used to determine the sequence of amino acids in a protein.

1495. translocation

Transfer of part of a chromosome from its normal location to a location on another chromosome.

1496. vector

A DNA carrier that can insert foreign genetic material into a host cell.

Evolution / Taxonomy

1497. abiogenesis

Beginning without life; spontaneous generation.

1498. adaptive radiation

Change in a population over time such as by divergence or convergence.

1499. adaptive value

Success of one genotype in a population relative to others.

1500. allopatric

Concerning species or populations separated by geographic barriers.

1501. analogous

Referring to structures having like functions but not necessarily the same embryonic precursors.

1502. anthropoid

An advanced primate, such as a monkey, ape, or human.

1503. artificial selection

Systematic and selective breeding directed by humans to bring out desired traits.

1504. balanced polymorphism

Equilibrium of homozygotes and heterozygotes maintained by natural selection.

1505. binomial system

The two-name system of naming organisms.

1506. biogenesis

Generating life from life.

1507. biological evolution

Changes over time in living organisms.

1508. character displacement

Increase in genetic difference between two populations or very similar species in the same geographic region.

1509. chemical evolution

The gradual increase in complexity of molecules thought to have preceded the origin of living cells.

1510. cladistics

Taxonomic classification using points of divergence in evolutionary lines.

1511. cladogenesis

Divergent evolution from common ancestors.

1512. class

Unit of classification under phylum and containing orders.

1513. coacervate droplet

A mixture of large molecules thought to have preceded the organization of the first cells.

1514. coevolution

Evolution in which adaptations of two lineages strongly affect each other.

1515. common ancestor

An ancestor to two or more branches in the evolutionary tree.

1516. comparative anatomy

The study of similarities and differences in structure among organisms.

1517. dichotomous key

A means of identifying organisms by chosing which of paired statements in a series pertain to an organism.

1518. directional selection

Selection involving changes that occur when a population displays a steady trend over time.

1519. disruptive selection

Selection due to unusual features that have high survival value.

1520. divergence

Radiating out in different directions.

1521. ecological equivalent

An unrelated organism having functions similar to another in the same environment.

1522. evolution

The process of change over time.

1523. evolutionary tree

A diagram showing evolutionary relationships among selected organisms.

1524. extinct

No longer having living representatives.

1525. family

Taxonomic unit within an order that includes related genera.

1526. fitness

Measure of evolutionary success in terms of relative number of genes contributed to gene pool.

1527. five kingdom system

A taxonomic system that places all living organisms in one of five kingdoms.

1528. fossil

Any evidence of organisms that lived in the past.

1529. founder effect

Evolutionary effect of a small, non-representative population giving rise to a species; extreme case of genetic drift.

1530. gene flow

The movement of genes from one population to another by reproduction between members of the populations.

1531. gene pool

The sum of all genes and their alleles present in a population at a given time.

1532. genetic drift

Fluctuations in gene frequencies due to isolation of non-representative sampling of the founding population.

1533. genus

The taxonomic category that combines similar species; the first part of the scientific name of an organism.

1534. Hardy-Weinberg equilibrium

A state in which gene frequencies remain unchanged from generation to generation in a population.

1535. hominid

A modern human or an ancestor.

1536. inclusive fitness

Fitness attributed to characteristics possessed by ancestors and close relatives.

1537. individual fitness

The extent to which an individual's genes survive in its offspring.

1538. isolating mechanism

A factor that prevents matings between members of two populations.

1539. kin selection

Tendency to protect close relatives, thereby fostering survival of one's own genes in others.

1540. kingdom

A major taxonomic subdivision of living organisms.

1541. labyrinthodont

Member of an extinct group of amphibians that probably gave rise to most other amphibians.

1542. macroevolution

Large scale evolutionary change.

1543. microevolution

Species-level or lower-level evolution; changes in gene frequencies.

1544. microsphere

A structure consisting of protein but having certain attributes of a cell.

1545. monophyletic

Evolved from the same ancestor.

1546. nested set

Hierarchical series of groups within groups, as in Linnaean classification.

1547. ontogeny

Developmental history of an organism.

1548. order

A taxonomic unit within class and containing families.

1549. orthogenesis

Evolution that proceeds in a straight line.

1550. oxidizing atmosphere

An atmosphere containing oxygen and other oxidizing molecules.

1551. pangea

Supercontinent of fused world land mass.

1552. parallel evolution

Independent evolution of similar structures in two species.

1553. paraphyletic

Concerning a taxonomic group that contains ancestors and some of their descendants.

1554. phyletic evolution

Evolution of a species directly along a linear path.

1555. phylogenetic tree

A branching diagram showing evolutionary relationships among selected organisms.

1556. phylogeny

Macroevolutionary history.

1557. phylum

Taxonomic unit within kingdom and including classes.

1558. polymorphic

Pertaining to a species having two or more distinct kinds of individuals.

1559. polyphyletic

Pertaining to a group of organisms classified together but having evolved from different ancestors.

1560. population

A set of all interacting, interbreeding individuals of one species.

1561. primitive atmosphere

Atmosphere having gases that were present prior to the emergence of life on earth.

1562. punctated equilibrium

Short period of speciation followed by a long period of little evolutionary change.

1563. race

An interbreeding population with particular gene frequencies different from those of other such groups.

1564. recapitulation

Tendency of embryos to resemble sequence of stages through which their ancestors evolved.

1565. scientific name

The genus and species to which an organism belongs.

1566. serial homology

Resemblance of parts of different organisms derived from a common ancestral source.

1567. speciation

The creation of a new species.

1568. species

A group of similar organisms having common genes.

1569. stasis

Lack of significant evolutionary changes in a phylogenetic line over a long time period.

1570. sympatric speciation

Origin of a new species from a group of an existing species because of physiological or behavioral isolation.

1571. synaptomorphy

Homologous structure found in two or more decendants of a common ancestor.

1572. systemics

Study of and classification according to evolutionary relationships.

1573. taxon

Unit of classification or group of organisms in it.

1574. taxonomic category

Level in a taxonomic hierarchy, such as order or family.

1575. taxonomy

The science of classifying organisms.

1576. vestigial

Concerning a structure that was once functional in an ancestor.

Energy / Populations

1577. aestivation

Dormancy of some animals during hot, dry season.

1578. amensalism

Relationship in which one organism benefits and the other is unaffected.

1579. biological control

The use of one organism to control another, especially the reduction in numbers of an undesirable organism.

1580. biomass

The total mass of all organisms living in a particular location.

1581. biotic potential

The maximum population growth rate under ideal conditions.

1582. browser

An animal that feeds above the ground on trees and shrubs.

1583. carrying capacity

The number of individuals of a species a particular environment can support indefinitely.

1584. climax community

The final community in ecological succession.

1585. cline

Continuous ranges of differences in structure and function extending over and related to geographic range.

1586. commensalism

The relationship of two species in which one benefits and the other is neither benefitted nor harmed.

1587. competitive exclusion principle

The principle that if two species continue to compete for the exact same resources, one will become extinct.

1588. consumer

An organism of one population that feeds on organisms of other populations within an ecosystem.

1589. cryptobiosis

Suspension of activity and metabolism during unfavorable environmental conditions.

1590. deciduous

The quality of losing leaves in winter or drought.

1591. decomposer

An organism that feeds on the remains of other organisms within an ecosystem.

1592. definitive host

Host to the sexually mature stage of a parasite.

1593. deme

Population of very similar, interbreeding organisms in a defined natural area.

1594. density-dependent factor

A population control factor that has a greater effect as the population size increases.

1595. density-independent factor

A population control factor that has the same effect regardless of the size of the population.

1596. deposit feeding

Feeding from deposits of detritis and minerals.

1597. detritus

Nonliving organic matter.

1598. doubling time

Time required for a population to double in size.

1599. ecological niche

The position of an organism in its environment; its diet, predators, habitat, and effects on its environment.

1600. emergent tree

A tree that stands taller than the others in a forest.

1601. environmental resistance

The carrying capacity of an environment for a species, stated as environment's resistance to population growth.

1602. euryhaline

Ability to tolerate a variety of salinities in environment.

1603. exponential growth

Growth that frequently doubles the population size.

1604. filter feeder

An animal that removes particulate food from water.

1605. food chain

The flow of energy and matter from the environment through organisms within an environment.

1606. food pyramid

A way of depicting the dependence of animals on other organisms in an ecosystem.

1607. food web

A pattern of interconnected food chains.

1608. geometric population growth

Increase in population by a constant percentage in each generation.

1609. grazer

An animal that feeds on grasses and other plants near the ground.

1610. humus

Partially decomposed remains of organisms and their wastes.

1611. irruptive growth

Population growth characterized by exponential growth, followed by catastrophic population reductions.

1612. krill

Assortment or planktonic crustaceans usually eaten by baleen whales.

1613. logistic growth equation

Mathematic population growth model that is restricted by environmental resistance.

1614. mortality

Rate at which members of a population die.

1615. mutualism

A relationship between two organisms of different species that benefits both organisms.

1616. natality

The rate at which new individuals are produced in a population.

1617. nekton

Collectively, organisms that are active swimmers.

1618. net productivity

Amount of energy stored per square meter of land surface.

1619. niche

The role an organism plays in its ecosystem.

1620. oligotrophic

Pertaining to an environment having few nutrients available to organisms.

1621. osmoconformers

Marine organisms whose bodies take on the same salt concentration as the environment.

1622. parasitism

A symbotic relationship in which one organism lives at the expense of and does some damage to the host organism.

1623. phoresis

A symbiotic relationship in which one species is transported by another.

1624. photoperiodism

Physiological responses to light and darkness.

1625. phytoplankton

Small water-dwelling plantlike protists.

1626. pioneer species

The first organisms to become established in an area at the beginning of ecological succession.

1627. plankton

A collection of free-floating, aquatic organisms carried by fresh water or ocean currents.

1628. population density

The number of one kind of organism in a defined area.

1629. predation

The eating of one organism by another.

1630. prey

An organism eaten by a predator.

1631. primary consumer

An organism that consumes plant material.

1632. producer

An organism that can make organic nutrients from inorganic substances in the environment.

1633. productivity

Rate of accumulation of organic matter by producers or of energy storage by consumers.

1634. replacement reproduction

A reproductive rate that replaces individuals that die and maintains a constant population size.

1635. salinity

Quantity of dissolved salts in a solution, in parts per thousand, for example.

1636. secondary consumer

An animal that eats animals that have fed on plants.

1637. seral stage

Any of the stages of ecological succession in an ecosystem.

1638. sere

Any stage in ecological succession of an ecosystem.

1639. stenohaline

Property that allows animals to live in only a restricted range of environmental salt concentrations.

1640. stratification

The layering of subcommunities as in a soil layers.

1641. substrate (of an organism)

Surface on which an organisms lives.

1642. succession

A series of ecological stages by which the community in a particular area gradually changes.

1643. symbiosis

A relationship between two interacting species of organisms.

1644. trophallaxis

Mutual exchange of food and secretions among members of an insect colony.

1645. trophic level

A categorization of species in a food web according to how they obtain nutrients.

1646. zonation

Divisions of a natural community according to variations in physical conditions.

1647. zooplankton

Animal-like protists usually floating in a body of water.

Ecosystems / Applications

1648. acid rain

Rain containing acids from atmospheric pollution.

1649. ammonification

Conversion of protein to ammonia by soil bacteria.

1650. benthos

Organisms found at the bottom of oceans or lakes.

1651. biodegradable

Capable of being decomposed by living organisms.

1652. biogeochemical cycle

Continuous passage of a mineral or other material in an ecosystem.

1653. biological magnification

The concentration of substances in tissues as they are passed up the food chain.

1654. biome

A major area of the earth having certain life forms maintained by the climate of the region.

1655. carbon cycle

A repetitive sequence of chemical processes in which carbon enters and leaves living organisms.

1656. chapparal

A biome containing broad leaved evergreen shrubs in a dense thicket.

1657. contour cultivation

The planting of crops across a slope to slow water runoff; terracing.

1658. coral reef

A structure in tropical waters formed by coral skeletons and proving shelter for many species.

1659. crop rotation

A system of changing the crop grown in a field to control pests and improve soil fertility.

1660. denitrification

The process of converting nitrate to nitrogen gas; a part of the nitrogen cycle.

1661. desalinization

Removal of salt, typically from ocean water.

1662. desert

An area receiving 25 cm or less of rain per year.

1663. desertification

The process by which an area of marginally useful farm land becomes a desert by overgrazing or other abuses.

1664. ecotone

Transitional region between biomes containing its own organisms and some from adjacent biomes.

1665. environmental impact statement

A statement of the findings of a detailed study of how an activity might affect the environment.

1666. epilimnion

Surface water in a lake or ocean.

1667. estuary

A body of water containing a mixture of fresh and salt water.

1668. euphotic zone

The surface layer of a body of water to the depth penetrated by light; region in which photosynthesis occurs.

1669. eutrophication

The process of aging and death of organisms in a pond or lake.

1670. fossil fuel

Combustible material, such as coal, oil and gas, derived from previously living material.

1671. grassland

A biome occupied mainly by grasses and having a dry climate.

1672. greenhouse effect

An increase in environmental temperature as carbon dioxide absorbs heat radiated from the earth.

1673. groundwater

Water in soil and in aquifers beneath the soil.

1674. habitat

The place where an organism normally lives.

1675. hazardous waste

Environmental pollutants that endanger living things.

1676. hydrothermal vent

A volcanic opening on the ocean floor from separation of tectonic plates that spews forth hot water and minerals.

1677. lentic

Pertaining to a lake or pond.

1678. littoral

Region of shallow water near the shore between high and low tides.

1679. lotic

Pertaining to a creek or river (running water) environment.

1680. mariculture

Growing of human food in the ocean.

1681. marine

Pertaining to the ocean.

1682. monoculture

The growing of only one species of crop over a large area.

1683. nitrification

Formation of nitrates by bacteria.

1684. nitrogen cycle

A sequence of repeated reactions that move nitrogen between organisms and their environment.

1685. nitrogen fixation

A process that chemically combines atmospheric nitrogen with other elements.

1686. nuclear waste

Unused radioactive material from nuclear power plants and other processes involving radiation.

1687. ozone shield

A layer of ozone in the upper atmosphere that protects the earth from damaging ultraviolet radiation.

1688. pelagic

Pertaining to organisms that inhabit open water.

1689. permafrost

Perennial frozen subsoil in Arctic or subarctic regions.

1690. pollution

The presence of a substance that damages the environment.

1691. prairie

A temperate grassland biome having rainfall between 25 and 40 cm per year.

1692. primary treatment

The first treatment given to sewage.

1693. salinization

Deposition of salt.

1694. salt marsh

Coastal grassland that undergoes seasonal flooding.

1695. savanna

A grassland biome that has occasional trees especially in Africa.

1696. secondary treatment

Bacterial decomposition in a sewage treatment plant.

1697. siltation

Deposition of silt.

1698. slash-and-burn agriculture

The practice of cutting and burning trees to prepare land for agriculture.

1699. strip-cropping

The planting of alternating strips of different crops in which one crop protects the other.

1700. taiga

A northern forest containing mainly conifer trees.

1701. temperate deciduous forest

A biome with a moderate climate and many large trees that lose their leaves.

1702. temperature inversion

An event that occurs when air near the ground is cooler than air above it.

1703. terracing

Farming on banks of soil built across a slope to reduce erosion; contour cultivation.

1704. tertiary treatment

A process that removes toxic substances and excess minerals from sewage effluent.

1705. thermal pollution

Abnormally high temperature produced in an environment.

1706. thermocline

A sharp temperature difference between layers in a body of water that prevents distribution of oxygen and nutrients.

1707. tropical rain forest

A hot moist environment with a very large number of species.

1708. tundra

A cold biome with permafrost and lacking trees.

1709. water cycle

A sequence of repetitive reactions in which water moves to and from living things.

1710. water table

A level below which aquifers are filled with water.

1711. wetland

Swamp or marsh that has standing water most of the year.

Behavior

1712. aggression

An attitude of hostility, which may or may not include an attack.

1713. agonistic behavior

Threatening action of one animal against another.

1714. altruism

Behavior that benefits others to the possible harm of the actor.

1715. anadromous

Concerning fish that migrate from oceans to fresh water to spawn.

1716. Batesian mimicry

A kind of mimicry in which a palatable species is protected from predators by similarity to an unpalatable one.

1717. biological clock

A mechanism that enables organisms to respond to time related changes in the environment.

1718. caste

A group of individual in a colony that performs particular tasks, such as workers bees.

1719. circadian rhythm

A biological cycle that is repeated about every 24 hours.

1720. cleaning symbiosis

A mutualistic relationship in which both organisms gain, one by cleaning and the other by being cleaned.

1721. competition

Interactions in which two organisms or two species limit each other's supply of food, shelter, or mates.

1722. conditioning

A kind of learning in which a stimulus and response become associated.

1723. courtship ritual

A pattern of activities preceding mating that are needed for mating to occur.

1724. dominance hierarchy

A social pattern in which a few animals dominate the majority.

1725. emotion

A state of feeling; the affective aspect of consciousness.

1726. engram

Memory trace.

1727. entrainment

The following of environmental cues in a time-related cycle.

1728. estivation

A period of inactivity during hot weather.

1729. ethology

The study of animal behavior.

1730. habit-forming

A property that causes some people to make great effort to obtain a drug.

1731. hibernation

A period of greatly reduced metabolism during cold weather.

1732. home range

The region to which an animal or a small group of animals normally confines its activities.

1733. imprinting

A special kind of learning that occurs during a brief sensitive time early in an animal's life.

1734. individual space

Space immediately surrounding an animal that it usually does not permit another animal to enter.

1735. innate

Already present at birth.

1736. latent learning

Learning having no immediate and apparent use.

1737. learned behavior

Behavior that results from experience.

1738. learning

A behavioral change in response to external stimuli.

1739. lek

Region where mating members of a species meet for courtship.

1740. memory

A process of storing and recalling previous experiences.

1741. mimicry

A resemblance of one organism to another that can defend it against a common predator.

1742. Mullerian mimicry

Resemblance among two or more unpalatable species.

1743. operant conditoning

Learning in which an animal is encouraged to repeat a behavior by being rewarded for performing it.

1744. pair bond

A special relationship between mating pairs that persists beyond the act of mating; seen in some birds and mammals.

1745. peck order

Hierarchy of dominance and recessivity in a social group (usually pertaining to birds).

1746. pheromone

A substance released by some organisms into the environment to communicate with others of the same species.

1747. physiological dependence

The need for a drug to prevent withdrawal symptoms.

1748. queen

A single mother of a colony of certain insects, such as bees.

1749. range

Region of earth over which a species is found.

1750. releaser

A stimulus that initiates a given behavior.

1751. social behavior

Communication and other behavioral interactions between two or more individuals.

1752. social hierarchy

Structure in a social group based on rank and usually maintained by aggressive behavior.

1753. sociobiology

The study of the biological basis of behavior.

1754. stereotyped behavior

Repetitive behavior that tends to follow a particular stimulus.

1755. stridulation

Audible vibrations produced by insects rubbing appendages together.

1756. taxis

Movement of a protist or animal toward or away from a stimulus.

1757. territoriality

Establishment and defense of a nesting, breeding, or feeding site in which others of the same species are driven away.

1758. territory

An area actively defended by an individual or group that inhabits it.

1759. tolerance

The requirement for larger doses of a substance to produce an effect.

1760. troop

Social unit of many primate species.

1761. warning coloration

Adaptation in which an organism without defense resembles a poisonous or unpalatable one.

1762. worker

A female insect that has no reproductive organs.

1763. zeitgeber

An environmental stimulus that serves to set an organism's biological clock.

Non-Arthropod Invertebrates

1764. acoelomate

Lacking a coelom (body cavity).

1765. ambulacral

Concerning radial grooves from which tube feet project.

1766. amebocyte

Wandering cell in a metazoan.

1767. apical complex

A group of organelles characteristic of certain protozoa.

1768. asconoid

A simple sponge with water canals that follow a straight course.

1769. bipinnaria

Ciliated, bilateral larval stage of an echinoderm.

1770. bivalve

Mollusks with two shells hinged dorsally.

1771. byssus

Threadlike secretion by which certain bivalve mollusks are anchored to substratum.

1772. cercaria

Free-swimming larva of a trematode parasite.

1773. chelicera

Pincerlike head appendages on many arachnids.

1774. choanocyte

Flagellated cell with thin cytoplasmic collar found in sponges and some protozoa.

1775. choanoderm

Layer of choanocytes in a sponge.

1776. clitellum

A dorsal saddle-like swelling that secretes a cocoon in certain segmented worms.

1777. cloaca

Common chamber for digestive, reproductive, and excretory discharges.

1778. cnida

Organelle that can be everted from a cnidocyte.

1779. cnidocyst

Part of a myxozoan spore containing a nematocyst-like organelle.

1780. cnidocyte

Cell that contains stinging structure (nematocyst).

1781. coelom

Body cavity; space between body wall and internal organs.

1782. coelomate

Having a body cavity fully lined with mesoderm.

1783. colony

Group of closely associated individuals usually formed by asexual budding.

1784. contractile vacuole

Organelle that regulates fluid balance in protozoans and sponges.

1785. ctenidium

Gill of a mollusk.

1786. cysticercus

An encysted developmental stage of a tapeworm.

1787. cytopharynx

Gullet-like organelle in certain protozoans.

1788. cytostome

Mouth-like structure in ciliates and some other protozoans.

1789. deuterostome

Member of a large group of animal phyla having an anus derived from a blastopore.

1790. dipleurula

A larval form unique to deuterostomes.

1791. endostyle

Longitudinal groove in the floor of the pharynx in primitive chordates that contains mucus-secreting ciliated cells.

1792. flame cell

A tubular network that constitutes the excretory system of a planarian.

1793. fluke

Common name for a group of trematodes (flat worms).

1794. incurrent pore

A pore through which water enters a sponge.

1795. invertebrate

An animal lacking a backbone.

1796. larva

Motile and sometimes feeding stage in the early development of many invertebrates.

1797. leucononid

Concerning a multibranched canal system in a sponge.

1798. lophophore

Ridge with ciliated tentacles around the mouth of some primitive animals.

1799. macronucleus

Large nucleus of a ciliate that controls activities other than reproduction.

1800. medusa

A free-floating inverted polyp seen in cnidarians.

1801. merozoite

Cell derived from splitting of a schizont of the malaria parasite.

1802. mesoglea

Nonliving jellylike substance separating ectoderm and endoderm in the sponge body wall.

1803. mesohyl

Middle region of sponge body that contains amoeboid cells and skeletal elements.

1804. metamerism

Body segmentation along its primary axis, producing a series of homologous parts.

1805. metazoan

A multicellular animal.

1806. micronucleus

Nucleus that controls reproduction in a ciliate.

1807. miracidium

First larval stage of a parasitic fluke.

1808. nacre

A layer of calcium carbonate that lines some mollusk shells.

1809. nematocyst

A threadlike group of stinging cells used by cnidarian to capture prey.

1810. opisthosoma

Posterior abdominal region in most arachnids.

1811. osculum

Excurrent pore in a sponge.

1812. pallium

Mantle, as in mollusks.

1813. paramylum

Carbohydrate stored by euglenoids.

1814. parapodium

A footlike fleshy lobe on the segments of marine annelids.

1815. pellicle

Surface covering of various protozoa.

1816. pinacoderm

Outer syncytial layer of cells in many sponges.

1817. plasmodium

Multinucleate cell in life cycle of many protozoa.

1818. polyp

A cnidarian body attached to a structure in its environment.

1819. proboscis

Tubular process usually used for feeding; snout.

1820. proglottid

A section of a tapeworm.

1821. protostome

One of many species of animals in which the mouth develops from the blastopore.

1822. pseudocoelom

A fluid-filled space between the mesoderm and the internal organs of an unsegmented roundworm.

1823. pygidium

Nonsegmented posterior part of a metameric organism.

1824. pyrenoid

Starch granule in chromatophores of some protozoa.

1825. radula

Rasp-like feeding structure of many mollusks.

1826. reticulum

Network of filaments.

1827. rhabdite

Rodlike organelle in epidermis of certain flatworms.

1828. schizocoel

Body cavity derived from splitting mesoderm into two layers.

1829. scolex

The head of a tapeworm.

1830. sessile

Remaining stationary.

1831. spherical symmetry

Symmetry in which cone-shaped sections are similar.

1832. spongin

Flexible protein fiber found in the skeleton of sponges.

1833. spongocoel

Main central water cavity in more complex sponges.

1834. strobila

Body consisting of a string of individuals; asexual stage of a cnidarian that produces medusae.

1835. style

Pointed rod as in the stomach of a clam.

1836. syconoid

Concerning a water canal system and body wall folding in certain sponges.

1837. tagma

Fused body segment, as head, thorax, or abdomen of an insect.

1838. tegument

Syncytial covering of parasitic flatworms.

1839. test

Protective case of many protozoa.

1840. trichocyst

Organelle that discharges a filament for trapping prey in cytoplasm of ciliated protozoa.

1841. trochophore

A free swimming, ciliated larva.

1842. tube foot

A hollow structure on the underside of a starfish used in food-getting and locomotion.

1843. tubicolous

Living inside a tube.

1844. veliger

Larva of many marine snails and clams.

1845. visceral mass

Soft internal organs of a mollusk.

1846. zoea

Early larval stage in many marine crustaceans.

1847. zooflagellate

Animal-like flagellated protozoan.

1848. zooxanthella

Dinoflagellate that lives symbiotically with coral or other marine animal.

Arthropods

1849. ametabolous

Lacking metamorphic changes.

1850. antenna

A sensory organ on the head of an arthropod.

1851. antennule

An organ of balance and hearing in some arthropods.

1852. arthropod

An animal with a segmented body and jointed appendages.

1853. carapace

Upper covering or shell on some arthropods.

1854. cephalothorax

The fused head and thorax of a crustacean.

1855. chelate

Having pincer-like appendages.

1856. chitin

A flexible substance found in exoskeletons.

1857. complete metamorphosis

Development in insects that proceeds through distinct egg, larva, pupa, and adult stages.

1858. compound eye

Arthropod eye with many functional units that produces multiple images.

1859. diapause

Inactive pupal state of an insect.

1860. ecdysis

Molting of arthropod skeleton.

1861. ecdysone

Hormone that induces molting in arthropods.

1862. endopodite

Inner branch of a biramous limb of a crustacean.

1863. exopodite

Outer branch of biramous limb of a crustacean.

1864. green gland

A gland that removes wastes from the blood in crayfish.

1865. haltere

A small second set of wings in certain kinds of flies.

1866. hemerythrin

Oxygen-binding protein in a few invertebrates in which two atoms of iron bind one oxygen molecule.

1867. hemimetabolous

Undergoing partial metamorphosis, as occurs in some insects.

1868. hemocoel

Blood filled body cavity in animals with an open circulatory system, such as arthropods.

1869. hemocyanin

An oxygen-carrying copper compound in the blood of mollusks.

1870. hemolymph

Fluid contained in open circulatory systems, which has properties of both blood and lymph.

1871. holometabolous

Undergoing complete metamorphosis.

1872. incomplete metamorphosis

Gradual development, as in some insects, from egg to nymph to adult.

1873. instar

Stage of arthropod between molts.

1874. jointed appendage

A movable, bendable structure extending from an animal's body.

1875. juvenile hormone

Hormone in arthropods that preserves juvenile characteristics during a molt.

1876. metamorphosis

Transformation of body structure, especially from larva to adult.

1877. molt-inhibiting hormone

Secretion of X gland that prevents molting of a crustacean.

1878. molting

Shedding of an exoskeleton.

1879. nauplius

Crustacean larva with three pairs of appendages.

1880. notum

Dorsal part of body; dorsal element of a segment in arthropods.

1881. nymph

An immature insect that differs little from one stage to the next.

1882. ocellus

Simple light-reflecting organ in invertebrates.

1883. ommatidium

A unit of a compound eye capable of forming an image.

1884. oviposter

A pointed organ in a female insect used to make a tunnel in the ground into which eggs are deposited.

1885. palps

Appendages in the mouth region used in feeding.

1886. pedipalp

A leglike sensory appendage found in spiders.

1887. pelopod

Biramous swimming appendage of a crustacean.

1888. pericardial sinus

A space surrounding the heart, especially in crayfish.

1889. peritrophic membrane

Chitin sheath secreted to enclose gut contents in insects.

1890. pleuron

Lateral skeletal piece of any segment of an arthropod.

1891. podium

A footlike structure.

1892. prosoma

Anterior body region in an arthropod or other invertebrate.

1893. protopodite

Basal part of biramous limb of a crustacean.

1894. pupa

Inactive, nonfeeding stage in insect development between the larva and adult.

1895. spinneret

An abdominal appendage that secretes silk in spiders.

1896. uropod

Abdominal segments in an arthropod that contribute to forming a paddle.

Vertebrates

1897. airfoil

Surface of a wing that produces lift.

1898. altricial

Pertaining to young that are born or hatch in helpless condition.

1899. ammocoete

Free swimming, filter-feeding larval stage (of a lamprey).

1900. amplexus

Clasping and pressing action of a male frog that releases eggs from the female's abdomen.

1901. aortic arch

Vertebrate embryonic blood vessel that branches from ventral to dorsal aorta.

1902. appendicular skeleton

Bones of girdles and limbs of vertebrates.

1903. arboreal

Tree-living.

1904. axial skeleton

Bones of cranium, vertebral column, and rib cage of vertebrates.

1905. carinate

Having a carina (keel).

1906. chordate

Animals that have a notochord at some stage of development.

1907. chromatophore

A pigmented cell found beneath the outer skin layer in some fishes.

1908. crop

A chamber in birds and earthworms where food is temporarily stored.

1909. ectotherm

An animal whose body temperature is determined by the environment.

1910. egg tooth

A single central tooth used by reptiles and birds during hatching from an egg and subsequently lost.

1911. eutherian

A kind of placental mammal with a well-formed placenta whose young are born relatively well developed.

1912. gizzard

A muscular chamber in birds and earthworms in which food is ground into small pieces by small stones.

1913. lagena

Evagination in vertebrate inner ear that gives rise to the cochlea.

1914. marsupial

Primitive mammal with a pouch in which extremely immature young continue development.

1915. notochord

A rigid rod of cells beneath the nerve cord that gives rise to vertebrae.

1916. pentadactyl

Having five digits on a limb.

1917. pineal eye

A light sensitive structure found in the brain of some extinct snakes.

1918. placoid scale

Dermal fish scale homologous with vertebrate tooth.

1919. precocial

Able to move and having hair or feathers soon after birth or hatching.

1920. primate

A mammal with grasping hands, well developed collar bones, and eyes directed forward.

1921. prosimian

The more primitive of two groups of primates.

1922. reptile

Terrestrial vertebrate having horny scales.

1923. retroperitoneal

Located behind the peritoneum.

1924. syrinx

Voice box in birds.

1925. teleost

Bony fish.

1926. tornaria

Free-swimming larva of hemichordate that closely resembles an echinoderm larva.

1927. vertebrate

Having a back bone.

1928. visceral arch

Cartilaginous or bony arch which forms walls of pharynx and supports gills in ancestral vertebrates.

Index

193

194

197

198

199

mesoglea	180	morphogenesis	136
mesohyl	180	mortality	162
mesolecithal	136	morula	136
mesonephros	82	mosaic	142
messenger RNA	147	motor end plate	48
metabolic rate	34	motor neuron	87
metabolism	34	motor unit	48
metacarpal	41	mucosa	55
metachronal rhythm	48	mucus	22
metamerism	180	Mullerian mimicry	174
metamorphosis	187	multiple alleles	142
metanephridium	82	muscle	22
metanephros	82	mutagen	148
metaphase	21	mutation	148
metaphasic chromosomes	21	mutualism	162
metastasis	21	myelencephalon	94
metatarsal	41	myelin	87
metazoan	180	myocardium	71
metencephalon	87	myofibril	48
meter	4	myofilament	49
micelle	54	myogenic	49
microevolution	155	myoglobin	49
microfilament	21	myomere	49
microglial cell	94	myoneural junction	49
micromere	136	myosin	49
micronucleus	180	myotome	49
micronutrient	60	nacre	181
microscope	4	nares	65
microsphere	155	natality	162
microtubule	21	natural selection	4
microvillus	54	naturally acquired active	
mictic	126	immunity	109
mimicry	174	naturally acquired passive	
mineral	60	immunity	109
mineralocorticoid	114	nauplius	187
miracidium	181	negative feedback	114
missense mutation	147	nekton	162
mitochondrion	21	nematocyst	181
mitosis	22	neocortex	94
mixed nerve	87	neonate	136
mixture	12	neoteny	136
mobility	48	nephridium	82
modiolus	100	nephron	82
molar	54	nephrostome	83
mole	12	nerve	87
molecular hybridization	147	nerve impulse	87
molecule	12	nerve net	88
molt-inhibiting hormone	187	nervous tissue	22
molting	187	nested set	155
monoculture	169	net filtration pressure	83
monocyte	78	net productivity	162
monoecious	127	net protein utilization	34
monohybrid	142	neural crest	136
monomer	13	neural oscillator	88
monophyletic	155	neural tube	136
monosaccharide	13	neurilemma	88
monosynaptic	87	neurite	88
monozygotic	136	neurofibrillary tangles	94
mons pubis	120	neurogenic	49

paramylum	181	phototrophic	34
paramyosin	50	phyletic evolution	156
paraphyletic	156	phylogenetic tree	156
parapodium	181	phylogeny	156
parasitism	163	phylum	156
parasympathetic division	114	physiological dependence	175
parathormone	114	physiology	5
parathyroid glands	114	phytoplankton	163
Parkinson's disease	94	pia mater	95
parthenogenesis	137	pinacoderm	181
partial pressure	65	pineal eye	191
parturition	120	pineal gland	95
passive immunity	110	pinna	101
passive transport	26	pinocytosis	22
patella	41	pioneer species	163
pattern generator	89	pitch	101
peck order	175	pituitary	115
pectoral	5	placenta	120
pedicel	41	placoid scale	192
pedipalp	188	plankton	163
pelagic	169	plantar	41
pellicle	181	planula	137
pelopod	188	plaque	71
pelvic	5	plasma	78
penetrance	143	plasma cell	110
penis	127	plasma membrane	27
pentadactyl	191	plasmin	78
pepsin	61	plasminogen	79
peptide bond	13	plasmodium	181
perception	95	plasmolysis	27
pericardial sinus	188	plasticity	42
pericardium	71	platelet	79
perikaryon	89	pleiotropy	143
perilymph	101	pleura	65
peripheral	5	pleuron	188
peripheral resistance	71	plexus	89
peristalsis	61	ploidy	143
peritoneum	55	podium	188
peritrophic membrane	188	poikilothermic	34
peritubular	83	point mutation	148
permafrost	169	polar body	128
permeability	27	polar compound	14
pernicious anemia	78	polarity	5
peroxisome	22	pollution	169
Peyer's patch	55	polyembryony	137
pH	14	polygenic inheritance	143
phagocytosis	110	polymer	14
phalanges	41	polymorphic	156
pharynx	65	polyp	182
phenocopy	143	polypeptide	14
phenotype	143	polyphyletic	156
pheromone	175	polyploidy	143
phoresis	163	polysaccharide	14
phospholipid	14	pons	95
phosphorescence	34	population	156
phosphorylation	34	population density	163
photon	101	portal system	71
photoperiodism	163	positive feedback	115
photoreceptor	101	posterior	5

postganglionic	89	punctated equilibrium	157
postovulatory	120	Punnett square	143
postsynaptic	89	pupa	189
potential energy	14	purine	148
prairie	170	Purkinje cell	95
precocial	192	Purkinje fiber	72
predation	163	pus	110
preganglionic	89	putrefaction	34
preovulatory	120	pygidium	182
prepuce	120	pylorus	55
pressure	66	pyrenoid	182
presynaptic	89	pyrimidine	148
prey	163	pyrogen	35
primary consumer	164	queen	175
primary follicle	128	race	157
primary treatment	170	radial cleavage	137
primate	192	radial symmetry	42
prime mover	50	radiation	15
primitive atmosphere	156	radius	42
primitive streak	137	radula	182
primordial	128	range	175
principle of dominance	143	raphe	42
principle of forward conduction	89	rapid-eye-movement (REM) sleep	95
principle of unit characters	143	rarefaction	102
proboscis	182	Rathke's pouch	115
producer	164	reabsorption	35
productivity	164	reactant	15
progesterone	115	recapitulation	157
proglottid	182	receptor	35
prokaryotic	22	recessive	144
prolactin	115	recombinant DNA	148
proliferative phase	120	rectum	55
pronephros	83	redia	137
prophase	23	reduction	15
proprioceptor	101	reflex	89
prosimian	192	refraction	102
prosoma	189	refractory period	89
prostaglandin	115	regeneration	137
prostate	128	regulator gene	148
prosthetic group	14	relaxin	115
protandry	128	releaser	175
protease	14	releasing hormone	115
protein	14	remission	23
proteoglycan	14	renal	83
prothrombin	79	renin	83
protogyny	128	renin-angiotensin mechanism	83
proton	15	rennin	61
protonephridium	83	replacement reproduction	164
protoplasm	23	replication	149
protopodite	189	repressor	149
protostome	182	reproduction	6
proximal	5	reproductive efficiency	128
pseudocoelom	182	reproductive engineering	121
pseudopodium	50	reptile	192
puberty	115	resistance	72
pubis	42	respiration	66
pulmonary	66	respiratory center	66
pulp cavity	55	respiratory pigment	66
pulse	72	response	89

205

208

www.ingramcontent.com/pod-product-compliance
Lightning Source LLC
Chambersburg PA
CBHW081116170526
45165CB00008B/2459

* 9 7 8 1 4 4 9 5 1 2 5 0 7 *